THE
CARUS MATHEMATICAL MONOGRAPHS

Published by
THE MATHEMATICAL ASSOCIATION OF AMERICA

———

THE CARUS MATHEMATICAL MONOGRAPHS are an expression of the desire of Mrs. Mary Hegeler Carus, and of her son, Dr. Edward H. Carus, to contribute to the dissemination of mathematical knowledge by making accessible at nominal cost a series of expository presentations of the best thoughts and keenest researches in pure and applied mathematics. The publication of the first four of these monographs was made possible by a notable gift to the Mathematical Association of America by Mrs. Carus as sole trustee of the Edward C. Hegeler Trust Fund. The sales from these have resulted in the Carus Monograph Fund, and the Mathematical Association has used this as a revolving book fund to publish the fifth and sixth monographs.

The expositions of mathematical subjects which the monographs contain are set forth in a manner comprehensible not only to teachers and students specializing in mathematics, but also to scientific workers in other fields, and especially to the wide circle of thoughtful people who, having a moderate acquaintance with elementary mathematics, wish to extend their knowledge without prolonged and critical study of the mathematical journals and treatises. The scope of this series includes also historical and biographical monographs.

The Carus Mathematical Monographs

NUMBER SIX

FOURIER SERIES AND
ORTHOGONAL POLYNOMIALS

By

DUNHAM JACKSON

Professor of Mathematics, The University of Minnesota

Published by

THE MATHEMATICAL ASSOCIATION OF AMERICA

Composed, printed and bound by
The Collegiate Press
George Banta Company, Inc.
Menasha, Wisconsin

PREFACE

The underlying theme of this monograph is that the fundamental simplicity of the properties of orthogonal functions and the developments in series associated with them not only commends them to the attention of the student of pure mathematics, but also renders them inevitably important in the analysis of natural phenomena which lend themselves to mathematical description.

It is the essence of mathematics that it concerns itself with those relations which lie so deep in the nature of things that they recur in the most varied situations. This is particularly true, of course, of the rudimentary notions of arithmetic and geometry which have forced themselves on the attention of mankind since the earliest beginnings of thought. But with the advance of science and the accompanying extension of the range of phenomena subjected to quantitative discussion, more highly organized groups of concepts, gradually simplified by reduction to their essentials, have come to manifest themselves with similar persistence.

Among these are the formulations relating to the general analytical concept of orthogonality, and, in a more restricted field, the particular types of orthogonal systems discussed in the following chapters.

The choice of material is guided by two main lines of development, which are intimately associated from the beginning, but ultimately diverge in accordance with alternative principles of generalization.

For the standard method of solution of the "partial differential equations of mathematical physics" by means of orthogonal functions the Fourier series may be regarded as the prototype, corresponding to the use of rectangular coordinates for purposes of geometric representation. Different choices of coordinate system lead to the series of Legendre, Laplace, and Bessel, as well as others not considered here at all. The words *Fourier series* in the title of the book are intended to typify, if not adequately to describe, this phase of the discussion.

The reader whose primary interest is in the physical applications may find it most satisfactory to begin with Chapters IV and V, referring back to the earlier chapters as the need for such reference becomes apparent, and postponing the study of convergence for later consideration.

As *orthogonal polynomials*, under the second half of the title, the Legendre polynomials are in a sense the simplest, having the constant unity as weight function. Other weight functions correspond to the Fourier series of sines and cosines separately (in connection with an auxiliary change of variable), the derivatives of the Legendre polynomials, the more general polynomials of Jacobi, and those of Hermite and Laguerre. From comparison of these various types it is natural to proceed to the consideration of orthogonal polynomials with an arbitrary weight function.

Within the compass of the matter to be presented, the order of topics has not been governed by strict adherence to either of the courses of development thus outlined. Various departures from what might be regarded after the event as the most direct logical sequence have appeared expedient in the interest of compactness of expression or facility of comprehension. In case of conflict between these desiderata the latter has been regarded as of paramount importance.

For the reading of most of the book no specific preparation is required beyond a first course in the calculus. In Chapters VIII, IX, and X, and in some of the exercises on Chapter III, an acquaintance with fundamental facts about the Gamma and Beta functions is assumed. A certain amount of "mathematical maturity" is presupposed, or should be acquired in the course of the reading. The reader must be able to appreciate the general notion of *function*, apart from representation by a preconceived type of formula; he must be ready to think of the value of a definite integral as a numerical magnitude, subject to relations of greater and less, and only incidentally as the result of a particular process of calculation. Especially, he must be prepared to accept the word *orthogonal* as a term of mathematical analysis, on the basis of its definition, with a geometric association only in the remote background or temporarily in abeyance altogether, as he has already learned to speak of a *linear* equation or the *square* of a number without feeling obliged to visualize a geometric figure in connection with every occurrence of the words.

Under the circumstances, "rigor" in the sense of literal completeness of statement has been out of the question. It is hoped however that the reader who is familiar with the methods of rigorous analysis will be able without any difficulty to read between the lines the requisite supplementary specifications, and will find that what has actually been said is entirely accurate in the light of such interpretation.

A set of exercises relating to the various chapters, and intended for the most part to illustrate and extend the text rather than to serve for purposes of "drill," has been grouped together at the end of the book.

A bibliography of suggestions for supplementary

reading has also been placed at the end, and citations in
the text, when not otherwise given in detail, refer to the
items of this bibliography. A short list of references to
the bibliography is attached at the end of each chapter.
A few more specific references have been inserted in
footnotes.

This book is based on a course which I have given
at the University of Minnesota over a long period of
years. It would be impossible now to make acknowledg-
ment to all the individuals in successive generations of
students whose comments I have consciously or uncon-
sciously taken into account. I am grateful to the Carus
Monograph Committee, and in particular to its Chair-
man, Professor Saunders Mac Lane, for the most cordial
cooperation and for numerous illuminating suggestions.

DUNHAM JACKSON

THE UNIVERSITY OF MINNESOTA
April, 1941

TABLE OF CONTENTS

CONTENTS   xi

FOURIER SERIES AND
ORTHOGONAL POLYNOMIALS

CHAPTER I

FOURIER SERIES

1. Definition of Fourier series. A given function $f(x)$ can be represented, under hypotheses of considerable generality, by an infinite series in the form

$$(1) \quad \begin{aligned} f(x) = A_0 + a_1 \cos x + a_2 \cos 2x + \cdots \\ + b_1 \sin x + b_2 \sin 2x + \cdots . \end{aligned}$$

Such a series, when the coefficients are determined in the manner to be described below, is called a *Fourier series*.

Since each term is a periodic function with period 2π, the sum of the series necessarily has the same period. (A function $\phi(x)$ is said to have a constant a as a period if $\phi(x+a)$ is identically equal to $\phi(x)$, even though a may not be the smallest value for which a relation of this sort is satisfied. If a is a period, any integral multiple of a is also a period. In accordance with this definition, $\cos nx$ and $\sin nx$ have the period 2π, although they have also the smaller period $2\pi/n$.)

On the other hand, a Fourier series is sometimes useful for the representation of a given function in a single interval of length 2π, when the property of periodicity is of no concern except as it results incidentally from evaluation of the series outside the interval in which the function was originally defined.

The period 2π may be replaced by one of arbitrary length, as will be pointed out later, with no additional difficulty beyond a slight loss of simplicity in the formulas.

1

2. Orthogonality of sines and cosines. The determination of the coefficients depends on the evaluation of certain definite integrals involving the sines and cosines which enter into the terms of the series. In the first place, if n is an integer not zero,

$$(2) \qquad \int_{-\pi}^{\pi} \cos nx \, dx = 0, \qquad \int_{-\pi}^{\pi} \sin nx \, dx = 0.$$

The latter relation holds also for $n=0$; in the former, setting $n=0$ replaces the right-hand member by 2π.

Throughout the rest of this section, p and q will be understood to represent non-negative integers. Since

$$\cos px \cos qx = \tfrac{1}{2} \cos (p - q)x + \tfrac{1}{2} \cos (p + q)x$$

it follows by application of the above with $n = p - q$ and again with $n = p + q$ that

$$(3) \qquad \int_{-\pi}^{\pi} \cos px \cos qx \, dx = 0$$

when $p \neq q$. If $q = p \neq 0$, the integral of $\cos (p+q)x$ over the period interval is still zero, while the other term gives

$$\int_{-\pi}^{\pi} \cos^2 px \, dx = \pi.$$

Similarly, the identities

$$\sin px \sin qx = \tfrac{1}{2} \cos (p - q)x - \tfrac{1}{2} \cos (p + q)x,$$

$$\sin px \cos qx = \tfrac{1}{2} \sin (p - q)x + \tfrac{1}{2} \sin (p + q)x$$

give

$$(4) \qquad \int_{-\pi}^{\pi} \sin px \sin qx \, dx = 0, \qquad\qquad p \neq q,$$

$$\int_{-\pi}^{\pi} \sin^2 px \, dx = \pi, \qquad p \neq 0,$$

(5)
$$\int_{-\pi}^{\pi} \sin px \cos qx \, dx = 0,$$

the last relation holding whether p and q are the same or different.

The vanishing of the integrals in (3), (4) and (5) is expressed in words by saying that *any two of the functions* 1, cos x, cos $2x$, $\cdots$, sin x, sin $2x$, $\cdots$ *are orthogonal to each other over the interval* $(-\pi, \pi)$.

In general, two functions $u(x)$, $v(x)$ *are said to be* ORTHOGONAL *to each other over an interval* (a, b) if

$$\int_{a}^{b} u(x)v(x)dx = 0.$$

This concept of orthogonality, to be regarded as a characteristic of functional relationship purely on the basis of its definition, is in fact related, not obviously perhaps but through certain stages of interpretation and generalization, to that of perpendicularity in geometry.*

3. Determination of the coefficients. If it is assumed for purposes of formal calculation that the series can be integrated term by term, integration of (1) with the use of (2) gives

$$\int_{-\pi}^{\pi} f(x)dx = 2\pi A_0, \qquad A_0 = \frac{1}{2\pi} \int_{-\pi}^{\pi} f(x)dx,$$

each integral on the right, with the exception of that of

---

* See Ex. 5 among the exercises on Chapter VII at the end of the book.

the constant term, reducing to zero. For the determination of a_k when $k \neq 0$ let the identity (1) be multiplied through by $\cos kx$, and let the resulting expression for $f(x) \cos kx$ be integrated from $-\pi$ to π, still under the assumption that integration term by term is legitimate. Again each integral on the right reduces to zero, in consequence of (2), (3), and (5), with the exception of a single one, in this case the one containing $\cos^2 kx$, and it is found that

$$\int_{-\pi}^{\pi} f(x) \cos kx \, dx = a_k \int_{-\pi}^{\pi} \cos^2 kx \, dx = \pi a_k,$$

$$(6) \qquad a_k = \frac{1}{\pi} \int_{-\pi}^{\pi} f(x) \cos kx \, dx.$$

Similarly,

$$(7) \qquad b_k = \frac{1}{\pi} \int_{-\pi}^{\pi} f(x) \sin kx \, dx.$$

It will be noticed that the formula for a_k does not reduce to that for A_0 if k is set equal to 0. However, if $2A_0$ is denoted by a_0, this a_0 is given by (6) with $k = 0$. Henceforth a Fourier series will regularly be written in the form

$$(8) \qquad \frac{a_0}{2} + \sum_{k=1}^{\infty} (a_k \cos kx + b_k \sin kx),$$

with all the coefficients, including a_0, given by (6) and (7).

The above derivation of the coefficients has involved certain assumptions with regard to the integration of the infinite series. If $f(x)$ is any integrable function on the interval $(-\pi, \pi)$, however, a set of coefficients a_k, b_k can be defined outright by the formulas (6), (7), and the resulting series (8) can be studied on its own merits as to its convergence and its validity as a representation of $f(x)$. This point of view will be adopted from now on; a Fourier series (8), whether convergent or not, will be associated with every function $f(x)$ for which the integrals (6), (7) have a meaning.

(The word *integrable*, here and throughout the subsequent pages, refers to the existence of the definite integral in question as limit of a sum, or, in the case of an improper integral, to the existence of the limit by which convergence of the improper integral is defined,* not to the possibility of evaluating the integral explicitly in terms of familiar formulas.)

If $F(y)$ is a function of the variable y with period $2p$, where p is an arbitrary positive number, let

$$x = \pi y/p, \qquad y = px/\pi,$$

and let $F(y)$ as a function of x be denoted by $f(x)$. Then $f(x)$ has the period 2π in terms of x. If $f(x)$ is represented by a series of the form (8), this constitutes a representation of $F(y)$ in the form

$$F(y) = \frac{a_0}{2} + \sum_{k=1}^{\infty} \left(a_k \cos \frac{k\pi y}{p} + b_k \sin \frac{k\pi y}{p} \right),$$

and the formulas (6), (7) for the coefficients become

$$a_k = \frac{1}{p} \int_{-p}^{p} F(y) \cos \frac{k\pi y}{p} \, dy,$$

$$b_k = \frac{1}{p} \int_{-p}^{p} F(y) \sin \frac{k\pi y}{p} \, dy.$$

---

* Alternatively, the word may be understood as stipulating the existence of the integral according to the definition of Lebesgue.

The whole theory of Fourier series is potentially of a corresponding degree of generality. The discussion will be continued, however, in terms of the simpler formulas associated with the particular value $p=\pi$.

It is obvious from the interpretation of a definite integral as the area under a curve, and readily proved on the basis of the analytical definition, that *if a periodic function is integrated over a period interval, this interval can be replaced by any other of the same length without changing the value of the integral.* If $\phi(x)$ has the period 2π,

$$\int_a^{a+2\pi} \phi(x)dx = \int_b^{b+2\pi} \phi(x)dx$$

for all values of a and b. In connection with the Fourier series for a function of period 2π the integrals may be written as extended over the interval $(0, 2\pi)$ instead of $(-\pi, \pi)$, and the use of still other period intervals is equally permissible and sometimes essential.

4. Series of cosines and series of sines. It is obvious likewise, and still more easily proved analytically than the assertion of the last paragraph, that if $\phi(x)$ is an even function, one such that $\phi(-x) \equiv \phi(x)$, and if it is integrated over any interval $(-a, a)$ symmetric with respect to the origin,

$$\int_{-a}^a \phi(x)dx = 2\int_0^a \phi(x)dx.$$

By way of proof, let the integral from $-a$ to a be written as the sum of those from $-a$ to 0 and from 0 to a, and in the integral from $-a$ to 0 let $t=-x$; then

$$\int_{-a}^{0} \phi(x) dx = -\int_{a}^{0} \phi(-t) dt = \int_{0}^{a} \phi(-t) dt$$

$$= \int_{0}^{a} \phi(t) dt,$$

and the two parts into which the whole integral has been resolved are equal.

Similarly, if $\phi(x)$ is odd, i.e. if $\phi(-x) \equiv -\phi(x)$,

$$\int_{-a}^{a} \phi(x) dx = 0,$$

the integrals from $-a$ and 0 and from 0 to a being equal in magnitude and opposite in sign.

If $f(x)$ is an even function in the interval $(-\pi, \pi)$, the function $f(x) \cos kx$ is even and the function $f(x) \sin kx$ is odd, for each value of k. When the coefficients in the Fourier series for $f(x)$ are defined by (6) and (7),

$$(9) \qquad a_k = \frac{2}{\pi} \int_{0}^{\pi} f(x) \cos kx \, dx$$

and $b_k = 0$. *The Fourier series for $f(x)$ contains only cosine terms, and the coefficients are given by* (9).

If $f(x)$ is odd, the products $f(x) \cos kx$ and $f(x) \sin kx$ are odd and even respectively; *the Fourier series contains only sine terms, the coefficients being given by*

$$(10) \qquad b_k = \frac{2}{\pi} \int_{0}^{\pi} f(x) \sin kx \, dx.$$

The formulas (9) and (10) in themselves involve the values of the function $f(x)$ only in the interval $(0, \pi)$. Any function which is integrable from 0 to π can be formally represented in that interval, without any assumption in advance that it is even or odd or periodic or defined elsewhere at all, by a series of cosines with coefficients (9), and alternatively by a sine series with coefficients (10).

5. Examples. Let the formulas of the last section be applied to the function $f(x) \equiv x$ in the interval $(0, \pi)$.

For the corresponding cosine series,

$$a_k = \frac{2}{\pi} \int_0^\pi x \cos kx \, dx.$$

When $k = 0$, direct integration gives immediately $a_0 = \pi$. For $k > 0$ let integration by parts be used, with x and $\cos kx \, dx$ as factors; it is found that

$$\int_0^\pi x \cos kx \, dx = [(1/k)x \sin kx]_0^\pi - \frac{1}{k} \int_0^\pi \sin kx \, dx$$

$$= (1/k^2)[\cos k\pi - 1],$$

so that $a_k = 0$ when k is even, and $a_k = -4/(\pi k^2)$ when k is odd. The resulting series is

$$(11) \quad \frac{\pi}{2} - \frac{4}{\pi} \left[\cos x + \frac{\cos 3x}{3^2} + \frac{\cos 5x}{5^2} + \cdots \right].$$

For the sine series,

$$\int_0^\pi x \sin kx \, dx = [-(1/k)x \cos kx]_0^\pi + \frac{1}{k} \int_0^\pi \cos kx \, dx$$

$$= (-\pi/k) \cos k\pi,$$

$$b_k = \frac{2}{\pi} \int_0^\pi x \sin kx \, dx = (-1)^{k-1} \frac{2}{k},$$

so that the series has the form

$$(12) \qquad 2\left[\sin x - \frac{\sin 2x}{2} + \frac{\sin 3x}{3} - \frac{\sin 4x}{4} + \cdots \right].$$

Each of these series is in fact convergent to the value x throughout the interval $(0, \pi)$, except for the right-hand end point $x = \pi$ in the case of the sine series. This will be established by the convergence proof to be given in §10 below. (The manner of convergence of certain series closely related to these is illustrated by the graphs accompanying Exs. 3, 4, 5 on Chapter I at the end of the book.) If the truth of the assertion is accepted meanwhile for the sake of argument, some striking consequences are immediately apparent.

Each term of (11) is an even function of period 2π. If x is given any value outside the interval $(0, \pi)$, the series is the same, term by term, as for a corresponding value of x in the interval. Being convergent in $(0, \pi)$, it is therefore necessarily convergent for all values of x, to a sum which is itself an even function with the period 2π. The graph of the sum is obtained by reflecting the line segment $y = x$, $0 \leq x \leq \pi$, in the y-axis, and repeating the broken line thus obtained for $-\pi \leq x \leq \pi$ in successive intervals of length 2π to the right and to the left. The whole graph is a zigzag line having corners at abscissas 0, $\pm\pi$, $\pm 2\pi$, $\cdots$. Analytically, the sum is a continuous function whose derivative is discontinuous for the values of x indicated.

In (12), on the other hand, each term is odd, having still the period 2π. The series is term by term the same for any value of x outside the interval $(0, \pi)$ as for a corresponding value of x in that interval, or the same except for a reversal of sign throughout. Its convergence is therefore certain when it is known to be convergent in $(0, \pi)$, and its sum is odd and periodic. The graph is symmetric with respect to the origin, and the sum is simply x throughout the interval $-\pi < x < \pi$. The line segment representing the sum in this interval, and crossing the x-axis at its middle point, is then to be translated horizontally to the right and to the left for repetition in successive period intervals, to give the rest of the graph. The sum of the series this time is itself a discontinuous function with breaks at the points $x = \pm \pi$, $\pm 3\pi, \cdots$. For these particular values of x the series is obviously convergent to the value 0, since each term vanishes separately.

The sum of a Fourier series, even in some of the simplest (and most important) cases, may thus be a quite different sort of function from those ordinarily dealt with in elementary analysis. The study of such series has been largely instrumental in bringing into currency the modern notion of function, according to which y is a function of x if definite values of y are associated with specified values of x in any way whatever, without restriction to the special types of dependence which first became familiar.

Discontinuities in function or derivative of the sort just described are to be sure by no means peculiar to Fourier series. They are equally characteristic of the other types of series to be studied later in this volume, and can be illustrated, though somewhat artificially, by series of more elementary form.

For example, the binomial series

$$1 + \frac{1}{2}\,y - \frac{1}{2\cdot 4}\,y^2 + \frac{1\cdot 3}{2\cdot 4\cdot 6}\,y^3 - \frac{1\cdot 3\cdot 5}{2\cdot 4\cdot 6\cdot 8}\,y^4 + \cdots$$

represents the non-negative square root of $1+y$ for $-1 \leqq y \leqq 1$. (A complete proof of this assertion for the end points of the interval is not entirely elementary, to be sure, but the form of the series is familiar.) If $y = x^2 - 1$, the non-negative square root of $1 + y$ is x when x is positive, and $-x$ when x is negative: $[1 + (x^2 - 1)]^{1/2} = |x|$. The interval $-1 \leqq y \leqq 1$ corresponds to the interval $0 \leqq x^2 \leqq 2$, $-2^{1/2} \leqq x \leqq 2^{1/2}$. In this interval the series

$$1 + \frac{1}{2}(x^2 - 1) - \frac{1}{2 \cdot 4}(x^2 - 1)^2 + \frac{1 \cdot 3}{2 \cdot 4 \cdot 6}(x^2 - 1)^3$$

$$- \frac{1 \cdot 3 \cdot 5}{2 \cdot 4 \cdot 6 \cdot 8}(x^2 - 1)^4 + \cdots$$

represents the function $|x|$, the graph of which has a corner at the origin. The term-by-term derivative of this series (the expression $(x^2 - 1)^k$ with its coefficient being regarded as a single term) represents the value 1, as derivative of $|x|$, for $0 < x \leqq 2^{1/2}$, and -1 for $-2^{1/2} \leqq x < 0$, and obviously converges to the value 0, since all the terms vanish, for $x = 0$.

6. Magnitude of coefficients under special hypotheses. Let $f(x)$ be a function of period 2π which has a continuous first derivative for all values of x. In the integral defining the Fourier coefficient a_k, integration by parts gives

$$\pi a_k = \int_{-\pi}^{\pi} f(x) \cos kx \, dx$$

$$= [(1/k)f(x) \sin kx]_{-\pi}^{\pi} - \frac{1}{k} \int_{-\pi}^{\pi} f'(x) \sin kx \, dx.$$

The product $f(x) \sin kx$ vanishes at both ends of the interval. If M_1 is the maximum of $|f'(x)|$,

$$\left| \int_{-\pi}^{\pi} f'(x) \sin kx \, dx \right| \leqq \int_{-\pi}^{\pi} \left| f'(x) \sin kx \right| dx$$

$$\leqq \int_{-\pi}^{\pi} M_1 dx = 2\pi M_1;$$

apart from the question of formal demonstration, the fact that the absolute value of a definite integral can not exceed the integral of the absolute value of the integrand is evident from the interpretation of the integral in terms of area. So $|a_k| \leqq 2M_1/k$. Similarly, $|b_k| \leqq 2M_1/k$; the expression $f(x) \cos kx$ which enters into the calculation does not vanish in general for $x = \pm\pi$, but it does take on the same value at both ends of the interval. The coefficients approach zero as k becomes infinite, with at least the degree of rapidity indicated by the inequalities obtained.

Suppose now that $f(x)$ has a continuous second derivative, with M_2 as the maximum of $|f''(x)|$. By two successive integrations by parts, with attention to the periodicity of the functions involved,

$$\pi a_k = \int_{-\pi}^{\pi} f(x) \cos kx \, dx = -\frac{1}{k} \int_{-\pi}^{\pi} f'(x) \sin kx \, dx$$

$$= -\frac{1}{k^2} \int_{-\pi}^{\pi} f''(x) \cos kx \, dx,$$

and $|a_k| \leqq 2M_2/k^2$. In the same way, $|b_k| \leqq 2M_2/k^2$.

Under the conditions of the last paragraph it can be inferred immediately that the Fourier series is convergent. For

$$\left| a_k \cos kx + b_k \sin kx \right| \leq 4M_2/k^2,$$

and the right-hand member is the general term of a convergent series. But from the point of view of completeness of demonstration this is not the same as saying that the series converges *to the value* $f(x)$; if the sine terms were omitted the remaining series of cosines would be no less convergent, but would not represent $f(x)$ in general. A proof that the series converges to the desired value, with less restrictive assumptions concerning $f(x)$, will be given in §10.

It will be convenient later to have similar inequalities for the coefficients under somewhat different hypotheses. Let $f(x)$ be continuous and of period 2π, and let it be supposed that the interval $(-\pi, \pi)$ can be divided into a finite number of subintervals, in each of which $f(x)$ is linear. The graph of $f(x)$ in any one period interval is then made up of a finite number of straight line segments of finite slope joined end to end. Such a function will be called for brevity a *broken-line function*. Let the abscissas of the corners in the interior of $(-\pi, \pi)$ be $x_1, x_2, \cdots, x_{m-1}$, and for uniformity of notation in the next formulas let $x_0 = -\pi$, $x_m = \pi$ (whether these points are corners or not). Let λ_j be the constant value of $f'(x)$ in the interval (x_{j-1}, x_j), and let λ be the largest of the numbers $|\lambda_j|$. For the jth subinterval,

$$
\int_{x_{j-1}}^{x_j} f(x) \cos kx \, dx
$$
$$
= \left[(1/k)f(x) \sin kx \right]_{x_{j-1}}^{x_j} - \frac{1}{k} \int_{x_{j-1}}^{x_j} \lambda_j \sin kx \, dx
$$
$$
= (1/k)\left[f(x_j) \sin kx_j - f(x_{j-1}) \sin kx_{j-1} \right]
$$
$$
+ (\lambda_j/k^2)\left[\cos kx_j - \cos kx_{j-1} \right].
$$

When a summation is performed over the m subintervals,

$$
\sum_{j=1}^{m} \left[f(x_j) \sin kx_j - f(x_{j-1}) \sin kx_{j-1} \right]
$$
$$
= f(\pi) \sin k\pi - f(-\pi) \sin(-k\pi) = 0,
$$

while $\left| (\lambda_j/k^2) \left[\cos kx_j - \cos kx_{j-1} \right] \right| \leqq 2\lambda/k^2$, and

$$\left| \sum_{j=1}^{m} (\lambda_j/k^2) \left[\cos kx_j - \cos kx_{j-1} \right] \right| \leqq 2m\lambda/k^2.$$

Consequently

$$\left| a_k \right| = \frac{1}{\pi} \left| \sum_{j=1}^{m} \int_{x_{j-1}}^{x_j} f(x) \cos kx \, dx \right| \leqq \frac{2m\lambda}{\pi k^2}\,.$$

Similarly $\left| b_k \right| \leqq 2m\lambda/(\pi k^2)$; the difference $f(\pi) \cos k\pi$ $-f(-\pi) \cos (-k\pi)$ is zero by reason of the periodicity of $f(x)$ and $\cos kx$, although its terms do not in general vanish separately. *The Fourier coefficients a_k, b_k of a broken-line function are such that*

(13) $$\left| a_k \right| \leqq C/k^2, \qquad \left| b_k \right| \leqq C/k^2,$$

where C is independent of k.

An upper bound of the same order of magnitude would be obtained for the coefficients if $f(x)$ merely had a continuous second derivative in each subinterval, instead of being linear there, corners being still admitted at the abscissas x_j, but this more general conclusion will not be needed.

It is clear that if $f(x)$ has a continuous derivative of higher order, this hypothesis can be used to prove still more rapid approach of the coefficients to zero.

7. Riemann's theorem on limit of general coefficient. Let $f(x)$ now be any function which is integrable over the interval $(-\pi, \pi)$, not necessarily periodic or defined at all outside that interval, but subject only to the additional restriction that $[f(x)]^2$ also is integrable over $(-\pi, \pi)$. Let $s_n(x)$ be the partial sum of its Fourier series through terms of the nth order,

(14) $\quad s_n(x) = \dfrac{a_0}{2} + \displaystyle\sum_{k=1}^{n} (a_k \cos kx + b_k \sin kx).$

It follows from the definition of the coefficients that

$$\int_{-\pi}^{\pi} f(x)s_n(x)dx = \frac{a_0}{2} \int_{-\pi}^{\pi} f(x)dx$$

$$+ \sum_{k=1}^{n} \left[a_k \int_{-\pi}^{\pi} f(x) \cos kx\, dx + b_k \int_{-\pi}^{\pi} f(x) \sin kx\, dx \right]$$

$$= \frac{\pi a_0^2}{2} + \pi \sum_{k=1}^{n} (a_k^2 + b_k^2),$$

and from the integral relations in §2, on expansion of $[s_n(x)]^2$ and integration term by term, that

$$\int_{-\pi}^{\pi} [s_n(x)]^2 dx = \frac{\pi a_0^2}{2} + \pi \sum_{k=1}^{n} (a_k^2 + b_k^2).$$

Consequently

$$\int_{-\pi}^{\pi} [f(x) - s_n(x)]^2 dx$$

$$= \int_{-\pi}^{\pi} [f(x)]^2 dx - 2 \int_{-\pi}^{\pi} f(x)s_n(x)dx + \int_{-\pi}^{\pi} [s_n(x)]^2 dx$$

$$= \int_{-\pi}^{\pi} [f(x)]^2 dx - \left[\frac{\pi a_0^2}{2} + \pi \sum_{k=1}^{n} (a_k^2 + b_k^2) \right].$$

The first member of this equality, being the integral of a square, is non-negative. It follows that

$$\frac{a_0^2}{2} + \sum_{k=1}^{n} (a_k^2 + b_k^2) \leqq \frac{1}{\pi} \int_{-\pi}^{\pi} [f(x)]^2 dx.$$

Since this is true for all values of n, while the right-hand side is independent of n, $\sum_1^\infty (a_k^2 + b_k^2)$ is convergent, and therefore, inasmuch as a necessary condition for the convergence of a series is that the general term approach zero,

$$(15) \qquad \lim_{k \to \infty} a_k = 0, \qquad \lim_{k \to \infty} b_k = 0.$$

The theorem that the Fourier coefficients approach zero, known as *Riemann's theorem*, is in fact true without the requirement that $[f(x)]^2$ be integrable. Proof of this more general fact will be omitted here. A corresponding proof for a closely related type of series, which can be readily modified (and simplified) so as to apply to Fourier series, is given in §4 of Chapter XI.

The substance of (15) as proved in the present section may be recorded in a different notation by saying that if $\phi(u)$ is a function (not necessarily periodic) such that both ϕ and ϕ^2 are integrable over $(-\pi, \pi)$,

$$(16) \qquad \begin{aligned} \lim_{n \to \infty} \int_{-\pi}^{\pi} \phi(u) \cos nu \, du &= 0, \\ \lim_{n \to \infty} \int_{-\pi}^{\pi} \phi(u) \sin nu \, du &= 0. \end{aligned}$$

A corollary, not particularly noteworthy in itself, will be used presently as a lemma. If the hypotheses are satisfied by $\phi(u)$ they will also be satisfied by the functions $\phi(u) \sin \frac{1}{2}u$ and $\phi(u) \cos \frac{1}{2}u$. Let the first of these products be substituted for $\phi(u)$ in the first equation of (16), and the other in the second equation. By addition of the results,

$$(17) \qquad \lim_{n \to \infty} \int_{-\pi}^{\pi} \phi(u) \sin (n + \tfrac{1}{2})u \, du = 0.$$

8. Evaluation of a sum of cosines. Further study of the convergence of Fourier series is based on a trigonometric identity. Let the sum

$$G(v) = \tfrac{1}{2} + \sum_{k=1}^{n} \cos kv$$

be multiplied by $2 \sin \tfrac{1}{2}v$. For $k \geqq 1$ let the products be evaluated by the relation

$$2 \sin \tfrac{1}{2}v \cos kv = \sin (k + \tfrac{1}{2})v - \sin (k - \tfrac{1}{2})v.$$

Thus

$$2 \sin \tfrac{1}{2}v\, G(v) = \sin \tfrac{1}{2}v + \sum_{k=1}^{n} \left[\sin (k + \tfrac{1}{2})v - \sin (k - \tfrac{1}{2})v \right]$$

$$= \sin (n + \tfrac{1}{2})v,$$

and

$$(18) \quad \tfrac{1}{2} + \cos v + \cos 2v + \cdots + \cos nv = \frac{\sin (n + \tfrac{1}{2})v}{2 \sin \tfrac{1}{2}v}.$$

9. Integral formula for partial sum of Fourier series. In $s_n(x)$ as defined by (14) let the formulas for the coefficients be written with t as variable of integration:

$$a_k = \frac{1}{\pi} \int_{-\pi}^{\pi} f(t) \cos kt\, dt, \qquad b_k = \frac{1}{\pi} \int_{-\pi}^{\pi} f(t) \sin kt\, dt.$$

In forming the products $a_k \cos kx$, $b_k \sin kx$ with these expressions for the coefficients, the factors $\cos kx$, $\sin kx$, being constant with respect to the variable of integration, can be written inside the integral sign, and the pair of terms $a_k \cos kx + b_k \sin kx$ can be represented by

$$\frac{1}{\pi} \int_{-\pi}^{\pi} f(t) \cos kt \cos kx \, dt + \frac{1}{\pi} \int_{-\pi}^{\pi} f(t) \sin kt \sin kx \, dt$$

$$= \frac{1}{\pi} \int_{-\pi}^{\pi} f(t) \cos k(t - x) dt.$$

So $s_n(x)$ has the representation

$$\frac{1}{\pi} \int_{-\pi}^{\pi} f(t) \left[\frac{1}{2} + \sum_{k=1}^{n} \cos k(t - x) \right] dt,$$

which by (18), with $v = t - x$, is equivalent to

$$(19) \qquad s_n(x) = \frac{1}{\pi} \int_{-\pi}^{\pi} f(t) \frac{\sin (n + \frac{1}{2})(t - x)}{2 \sin \frac{1}{2}(t - x)} \, dt.$$

Let it be supposed now that $f(x)$ has the period 2π. In (19) let the variable of integration be changed by the substitution $u = t - x$. The limits of integration with respect to u are in the first instance $-\pi - x$ and $\pi - x$; by the observation in the last paragraph of §3, however, since the integrand has the period 2π with respect to u, the integral from $-\pi - x$ to $\pi - x$ has the same value as that from $-\pi$ to π. So

$$(20) \qquad s_n(x) = \frac{1}{\pi} \int_{-\pi}^{\pi} f(x + u) \frac{\sin (n + \frac{1}{2})u}{2 \sin \frac{1}{2}u} \, du.$$

(The periodicity of the fraction as a function of u is apparent from (18); in the fraction itself, increase of u by 2π reverses the algebraic signs of numerator and denominator separately, but leaves the value of the ratio unchanged.)

10. Convergence at a point of continuity. By integration of (18) from $-\pi$ to π,

$$(21) \qquad \int_{-\pi}^{\pi} \frac{\sin (n + \frac{1}{2})u}{2 \sin \frac{1}{2}u}\, du = \pi.$$

Let this equation be multiplied by $(1/\pi) f(x)$; since $f(x)$ is a constant with respect to the variable u, it can be written under the sign of integration:

$$(22) \qquad f(x) = \frac{1}{\pi} \int_{-\pi}^{\pi} f(x) \frac{\sin (n + \frac{1}{2})u}{2 \sin \frac{1}{2}u}\, du.$$

By subtraction of (22) from (20),

$$(23) \quad s_n(x) - f(x) = \frac{1}{\pi} \int_{-\pi}^{\pi} [f(x+u) - f(x)] \frac{\sin (n + \frac{1}{2})u}{2 \sin \frac{1}{2}u}\, du.$$

The proof of convergence consists in showing that under suitable hypotheses this expression approaches zero as n becomes infinite.

Let $f(x)$ be an integrable function of period 2π such that $[f(x)]^2$ also is integrable over a period. This condition will certainly be satisfied if $f(x)$ is everywhere continuous, or if it is continuous except for a finite number of finite jumps in a period, a finite jump being a point of discontinuity at which the function approaches a limit from the right and approaches a different limit from the left. Let attention be concentrated on the question of convergence at a specified point, and let it be assumed (for the present) that $f(x)$ is continuous at the point in question. The value of x being regarded as fixed, let

$$\phi(u) = \frac{f(x + u) - f(x)}{2 \sin \frac{1}{2}u}.$$

Then by (23)

$$(24) \qquad s_n(x) - f(x) = \frac{1}{\pi} \int_{-\pi}^{\pi} \phi(u) \sin (n + \frac{1}{2})u\, du.$$

The quotient $\phi(u)$ can be written in the form

$$\phi(u) = \frac{f(x+u) - f(x)}{u} \cdot \frac{\frac{1}{2}u}{\sin \frac{1}{2}u} .$$

The fraction $(\frac{1}{2}u)/\sin (\frac{1}{2}u)$ approaches the limit 1 as u approaches 0, and if defined by this limiting value for $u=0$ is continuous for $-\pi \leq u \leq \pi$. The condition that $[f(x+u)-f(x)]/u$ approach a limit for $u=0$ is precisely the condition that $f(t)$ have a derivative for $t=x$, by the definition of a derivative. If a derivative exists in the strict sense, i.e., if the difference quotient approaches the same limit as u approaches 0 from either side (the word *limit* referring always to a finite limit), $\phi(u)$ is continuous for $u=0$ if defined there by its limiting value. If different limits are approached from the right and from the left, as at a point where the graph of $f(t)$ has a corner, $\phi(u)$ has a finite jump for $u=0$. In either case, if $f(t)$ is continuous except for a finite number of finite jumps in a period, the same is true of $\phi(u)$ in the interval $(-\pi, \pi)$. By (17) then, as applied to (24),

$$\lim_{n \to \infty} [s_n(x) - f(x)] = 0.$$

The conclusion may be stated as follows:

If $f(x)$, having the period 2π, is everywhere continuous, or is continuous except for a finite number of finite jumps in a period, its Fourier series converges to the value $f(x)$ at every point of continuity where $f(x)$ has a right-hand and a left-hand derivative, whether these are the same or different.

The conclusion holds even if no derivative exists at the point, and without reference to the details of the last two paragraphs, provided that $[\phi(u)]^2$ is integrable. Still more generally, it is sufficient that $\phi(u)$ and $|\phi(u)|$

be integrable, if Riemann's theorem of §7 is assumed as known in the correspondingly more general form.

11. Uniform convergence under special hypotheses. The proof of the last section applies in particular if $f(x)$ is a broken-line function as described in §6. It was already apparent from the reasoning of §6 that the Fourier series for such a function is convergent; it is now assured that the sum of the series is $f(x)$ for all values of x.

Since the series actually represents $f(x)$, the difference between $f(x)$ and $s_n(x)$ can be written in the form

$$f(x) - s_n(x) = \sum_{k=n+1}^{\infty} (a_k \cos kx + b_k \sin kx).$$

In consequence of (13), $\left| a_k \cos kx + b_k \sin kx \right| \leq 2C/k^2$, and therefore

$$\left| f(x) - s_n(x) \right| \leq 2C \sum_{k=n+1}^{\infty} \frac{1}{k^2}.$$

Since $1/k^2 \leq 1/t^2$ for $k-1 \leq t \leq k$ it follows that

$$\frac{1}{k^2} = \int_{k-1}^{k} \frac{dt}{k^2} \leq \int_{k-1}^{k} \frac{dt}{t^2}, \qquad \sum_{k=n+1}^{\infty} \frac{1}{k^2} \leq \int_{n}^{\infty} \frac{dt}{t^2} = \frac{1}{n}.$$

Hence, for all values of x,

$$\left| f(x) - s_n(x) \right| \leq 2C/n.$$

The right-hand member is *independent of x*, and approaches zero as n becomes infinite. The fact that the remainder satisfies a relation of this sort is expressed by saying that the series is *uniformly convergent*:

The Fourier series for a broken-line function (of the type specified) *converges uniformly to the function for all values of x.*

A similar statement holds on the basis of §§6 and 10

for a periodic function which has a continuous second derivative everywhere. A proof of uniform convergence under more general hypotheses will be given in §21.

The discussion of the present section of course applies in particular to the cosine series (11) of §5, regarded as the Fourier series for the function of period 2π which is equal to $|x|$ for $-\pi \leqq x \leqq \pi$, and justifies the statement made in §5 about the convergence of that series.

12. Convergence at a point of discontinuity. The sine series (12) in §5 may similarly be regarded as the Fourier series for the discontinuous periodic function there described, and the proof of §10 establishes its convergence except for the isolated values of x at which the discontinuities occur. At these points, as noted in §5, the convergence of the particular series in question is obvious, its sum being 0. The significant thing is that the series converges to the value half-way between the limits approached by the function as the discontinuity is approached from the right and from the left. It is to be shown that this behavior of the series at a finite jump is typical.

Let $f(x)$ be a function of period 2π, assumed for simplicity to be continuous except for a finite number of finite jumps in a period. Let $f(x+)$ and $f(x-)$ denote the limits approached from the right and from the left at the point x, equal or different according as the point is one of continuity or of discontinuity. For given x let

$$\phi_1(u) = \frac{f(x+u) - f(x+)}{2 \sin \frac{1}{2}u},$$

$$\phi_2(u) = \frac{f(x+u) - f(x-)}{2 \sin \frac{1}{2}u},$$

ϕ_1 being defined for $u > 0$ and ϕ_2 for $u < 0$. Let it be supposed, for the sake of simplicity again rather than for the greatest generality, that each of the difference quotients

$$[f(x + u) - f(x+)]/u, \quad [f(x + u) - f(x-)]/u$$

approaches a limit as u approaches zero through values having the appropriate algebraic sign, i.e. that the function equal to $f(t)$ for $t > x$ and equal to $f(x+)$ for $t = x$ has a right-hand derivative at the point $t = x$, with a corresponding interpretation on the left. Then $\phi_1(u)$ and $\phi_2(u)$ likewise approach limits for $u = 0$, and if defined by their limiting values there are continuous except for a finite number of finite jumps throughout $(0, \pi)$ and $(-\pi, 0)$ respectively.

The hypotheses on which (17) is based are fulfilled if ϕ and ϕ^2 are integrable over $(0, \pi)$ and ϕ is identically zero on the interval $(-\pi, 0)$. So if ϕ is any function such that ϕ and ϕ^2 are integrable over $(0, \pi)$,

$$\lim_{n \to \infty} \int_0^\pi \phi(u) \sin (n + \tfrac{1}{2})u \, du = 0.$$

A similar observation applies to the interval $(-\pi, 0)$. Since the integrand in (21) is an even function of u,

$$\int_{-\pi}^0 \frac{\sin (n + \tfrac{1}{2})u}{2 \sin \tfrac{1}{2}u} \, du = \int_0^\pi \frac{\sin (n + \tfrac{1}{2})u}{2 \sin \tfrac{1}{2}u} \, du = \frac{\pi}{2}.$$

The integral in (20) may be regarded as the sum of integrals from $-\pi$ to 0 and from 0 to π. By steps analogous to those which led to (23) and (24),

$$\tfrac{1}{2}f(x +) = \frac{1}{\pi} \int_0^\pi f(x +) \frac{\sin (n + \tfrac{1}{2})u}{2 \sin \tfrac{1}{2}u} \, du,$$

$$\tfrac{1}{2}f(x-) = \frac{1}{\pi} \int_{-\pi}^{0} f(x-) \frac{\sin (n + \tfrac{1}{2})u}{2 \sin \tfrac{1}{2}u} \, du,$$

$$s_n(x) - \tfrac{1}{2}\big[f(x+) + f(x-)\big]$$

$$= \frac{1}{\pi} \int_{0}^{\pi} \phi_1(u) \sin (n + \tfrac{1}{2})u \, du$$

$$+ \frac{1}{\pi} \int_{-\pi}^{0} \phi_2(u) \sin (n + \tfrac{1}{2})u \, du.$$

Under the hypotheses imposed on $f(x)$ in the second paragraph of this section, each of the last two integrals approaches zero as n becomes infinite, and

$$\lim_{n \to \infty} s_n(x) = \tfrac{1}{2}\big[f(x+) + f(x-)\big].$$

The series converges to the mean of the limits approached by $f(x)$ from the right and from the left.

In consequence of a well known theorem of the theory of functions, the series can not converge uniformly in the neighborhood of a point of discontinuity. A striking peculiarity of the manner of convergence of the partial sums near a finite jump, involving, from the point of view of graphical representation, a careful distinction between variation of the ordinate for a fixed value or fixed interval of values of x, as n increases, and variation of the whole curve as a geometric figure in two dimensions, is known as the *Gibbs phenomenon*.*

13. Sufficiency of conditions relating to a restricted neighborhood. A further conclusion of a high degree of generality can be deduced from (24).

* See e.g. Bôcher, pp. 123–132; Zygmund, pp. 179–181. (Here and subsequently, when references are given in abbreviated form, details are to be supplied from the Bibliography at the end of the book.)

If $f(t)$ and $[f(t)]^2$ are integrable over a period, and if there is an interval $(x-h, x+h)$ throughout which $f(t)$ vanishes identically, then $\phi(u)$ and $[\phi(u)]^2$ are integrable from $-\pi$ to π, since the numerator of $\phi(u)$ is identically zero throughout the neighborhood of the point where the denominator vanishes. So $\lim_{n\to\infty} s_n(x) = f(x) = 0$.

Let $f_1(t)$ and $f_2(t)$ be two functions, each integrable with its square over a period, and let these functions be identically equal to each other throughout an interval $(x-h, x+h)$. Let $s_{n1}(x)$ and $s_{n2}(x)$ be the values of the partial sums of the Fourier series for f_1 and f_2 respectively at the point x. Then $s_{n1}(x) - s_{n2}(x)$ is the partial sum of the series for $f_1 - f_2$, and since this approaches zero, by the preceding paragraph, it follows that if one of the sums $s_{n1}(x)$, $s_{n2}(x)$ approaches a limit, the other approaches the same limit, regardless of differences between the functions f_1 and f_2 outside the interval $(x-h, x+h)$. The conclusion can be expressed by saying that (subject to the condition that the functions and their squares are integrable over a period) *the convergence of the Fourier series for a function at a specified point depends only on the behavior of the function in the neighborhood of the point.* The hypothesis of integrability of the squares can be dispensed with if Riemann's theorem of §7, on which the demonstration is based, is used in the more general form.

14. Weierstrass's theorem on trigonometric approximation. By a *trigonometric sum* is meant an expression of the form

$$\frac{\alpha_0}{2} + \alpha_1 \cos x + \alpha_2 \cos 2x + \cdots + \alpha_n \cos nx$$

$$+ \beta_1 \sin x + \beta_2 \sin 2x + \cdots + \beta_n \sin nx,$$

with any constant coefficients α_k, β_k. This sum is said to be *of the nth order* if α_n and β_n are not both zero. In many connections it is immaterial whether the terms in cos nx and sin nx are actually present or not, and the designation "trigonometric sum of the nth order" may then be used for brevity to mean a sum of the nth order *at most.*

A theorem of Weierstrass* states that *any continuous function f(x) of period 2π can be uniformly approximated by a trigonometric sum with any preassigned degree of accuracy.* That is to say, if ϵ is any positive quantity, there exists a trigonometric sum $T(x)$ (of some order) such that

$$(25) \qquad \left| f(x) - T(x) \right| < \epsilon$$

for all values of x.

If the Fourier series for $f(x)$ is uniformly convergent, the requirement of (25) is satisfied by taking for $T(x)$ any partial sum of the series of sufficiently high order. It can be shown, however, that *there are continuous functions which have divergent Fourier series.* While functions of this sort are complicated and artificial in structure, and will not be illustrated here, Weierstrass's theorem as stated derives significance from the fact that such functions exist. The sum $T(x)$ given by the demonstration of the theorem will not be merely a partial sum of the Fourier series for $f(x)$.

Let $f(x)$ then be a function of period 2π which is continuous everywhere, but not subject to any other restriction. Let ϵ be an arbitrary positive number. Let the interval $(-\pi, \pi)$ be divided into m parts, for sim-

* K. Weierstrass, *Über die analytische Darstellbarkeit sogenannter willkürlicher Functionen einer reellen Veränderlichen*, zweite Mittheilung, Berliner Sitzungsberichte, 1885, pp. 789–805.

plicity equal, by points $x_1, x_2, \cdots, x_{m-1}$, with $x_0 = -\pi$, $x_m = \pi$: $x_j = -\pi + 2j\pi/m$. Let $g(x)$ be the broken-line function of period 2π which takes on the same value as $f(x)$ at each of the points x_j, and is linear in each interval (x_j, x_{j+1}). It is clear that $g(x)$ can be made to differ from $f(x)$ by less than any preassigned amount by taking m sufficiently large. (This is an immediate consequence of fundamental properties of continuous functions, and may be regarded as obvious if the word *continuous* is taken as descriptive and self-explanatory rather than a matter of formal definition.) In particular it is possible to take m so large that

$$(26) \qquad \left| f(x) - g(x) \right| < \tfrac{1}{2}\epsilon$$

for all values of x. Let such a value of m be chosen and henceforth regarded as fixed, the function $g(x)$ being defined accordingly.

By §11 the Fourier series for $g(x)$ is uniformly convergent. Let $t_n(x)$ be the partial sum of this series through terms of the nth order. The uniform convergence means that if n is sufficiently large

$$(27) \qquad \left| g(x) - t_n(x) \right| < \tfrac{1}{2}\epsilon$$

for all x. When any such value is assigned to n,

$$\left| f(x) - t_n(x) \right| < \epsilon,$$

by combination of (26) and (27), and $t_n(x)$ will serve as the sum $T(x)$ in (25).

15. Least-square property. Let $f(x)$ be any integrable function with integrable square on the interval $(-\pi, \pi)$, and let $s_n(x)$ as in (14) be the partial sum of its Fourier series. Let $f(x) - s_n(x) = r_n(x)$. From the equations

$$\int_{-\pi}^{\pi} f(x) \cos kx \, dx = \pi a_k = \int_{-\pi}^{\pi} s_n(x) \cos kx \, dx,$$

$$k = 0, 1, \cdots, n,$$

and the corresponding pair involving b_k and $\sin kx$ it follows that

$$(28) \quad \int_{-\pi}^{\pi} r_n(x) \cos kx \, dx = \int_{-\pi}^{\pi} r_n(x) \sin kx \, dx = 0$$

for $k \leqq n$. Let

$$t_n(x) = \frac{\alpha_0}{2} + \sum_{k=1}^{n} (\alpha_k \cos kx + \beta_k \sin kx)$$

be an arbitrary trigonometric sum of the nth (or lower) order, and let $t_n(x) - s_n(x) = \rho_n(x)$, $\alpha_k - a_k = \gamma_k$, $\beta_k - b_k = \delta_k$, so that

$$\rho_n(x) = \frac{\gamma_0}{2} + \sum_{k=1}^{n} (\gamma_k \cos kx + \delta_k \sin kx),$$

$$f(x) - t_n(x) = r_n(x) - \rho_n(x).$$

As a consequence of (28).

$$\int_{-\pi}^{\pi} r_n(x)\rho_n(x)dx = 0,$$

while

$$\int_{-\pi}^{\pi} [\rho_n(x)]^2 dx = \frac{\pi\gamma_0^2}{2} + \pi \sum_{k=1}^{n} (\gamma_k^2 + \delta_k^2),$$

which is positive unless all the γ's and δ's are zero. So, if $t_n(x)$ is any trigonometric sum of the nth order at most, other than $s_n(x)$,

$$\int_{-\pi}^{\pi} [f(x) - t_n(x)]^2 dx = \int_{-\pi}^{\pi} [r_n(x) - \rho_n(x)]^2 dx$$

$$= \int_{-\pi}^{\pi} [r_n(x)]^2 dx - 2\int_{-\pi}^{\pi} r_n(x)\rho_n(x) dx$$

$$+ \int_{-\pi}^{\pi} [\rho_n(x)]^2 dx$$

$$= \int_{-\pi}^{\pi} [f(x) - s_n(x)]^2 dx + \int_{-\pi}^{\pi} [\rho_n(x)]^2 dx$$

$$> \int_{-\pi}^{\pi} [f(x) - s_n(x)]^2 dx.$$

Considered as an approximation to $f(x)$, the partial sum $s_n(x)$ of the Fourier series is distinguished among all trigonometric sums of the nth order at most as the one for which the integral of the square of the error is a minimum.

The same conclusion is indicated by regarding

$$\int_{-\pi}^{\pi} [f(x) - t_n(x)]^2 dx$$

for arbitrary $t_n(x)$ as a function of the $2n+1$ variables α_k, β_k, and setting its partial derivatives with respect to these variables equal to zero; but this procedure, unless supplemented by additional reasoning, naturally establishes only the necessity of the conditions $\alpha_k = a_k$, $\beta_k = b_k$ for a minimum, not their sufficiency.

16. Parseval's theorem. Let $f(x)$ again be a continuous function of period 2π, otherwise unrestricted, and $s_n(x)$ the partial sum of its Fourier series. Let η be an arbitrary positive number. According to §7,

$$(29) \quad \int_{-\pi}^{\pi} [f(x)]^2 dx - \left[\frac{\pi a_0^2}{2} + \pi \sum_{k=1}^{n} (a_k^2 + b_k^2) \right]$$

$$= \int_{-\pi}^{\pi} [f(x) - s_n(x)]^2 dx.$$

Let $\epsilon = [\eta/(2\pi)]^{1/2}$. By §14 there exists a trigonometric sum $t_n(x)$, of some order n, such that $|f(x) - t_n(x)| < \epsilon$, which makes

$$\int_{-\pi}^{\pi} [f(x) - t_n(x)]^2 dx < 2\pi\epsilon^2 = \eta;$$

and by §15, applied to the partial sum of corresponding index,

$$\int_{-\pi}^{\pi} [f(x) - s_n(x)]^2 dx \leqq \int_{-\pi}^{\pi} [f(x) - t_n(x)]^2 dx.$$

So for every positive η there exists an n such that the difference in the left-hand member of (29) is less than η. If this is true for a particular n it is true for every larger n, since the terms of the summation are non-negative. The difference therefore approaches zero as n becomes infinite, which means that

$$(30) \quad \frac{a_0^2}{2} + \sum_{k=1}^{\infty} (a_k^2 + b_k^2) = \frac{1}{\pi} \int_{-\pi}^{\pi} [f(x)]^2 dx.$$

The convergence of the series was already apparent in §7, together with the fact that its sum is not greater than the quantity on the right; the additional fact now established is that the relation is an equality for every continuous $f(x)$.

This conclusion, or a broader one expressing the

validity of (30) under still more general hypotheses with
regard to $f(x)$, is known as *Parseval's theorem*.

17. Summation of series. If $u_0 + u_1 + u_2 + \cdots$ is an
arbitrary infinite series, let

$$s_n = u_0 + u_1 + u_2 + \cdots + u_n.$$

Convergence means that s_n approaches a limit as n becomes infinite. It can happen that the quantity

$$\sigma_n = (s_0 + s_1 + \cdots + s_{n-1})/n,$$

the average of the first n partial sums, approaches a
limit even when the series does not converge. In the case
of the series

$$1 - 1 + 1 - 1 + \cdots,$$

for example, s_n takes on alternately the values 1 and 0,
and does not approach a limit, while σ_n has the value $\frac{1}{2}$
when n is even and the value $\frac{1}{2} + 1/(2n)$ when n is odd,
and approaches the limit $\frac{1}{2}$. If σ_n does have a limit A,
the series is said to be *summable by the method of the first
arithmetic mean to the value A*. This notion of summability is a generalization of that of convergence, in
the sense that a series which is not convergent may be
summable, as in the illustration just given, while *a series
which is convergent is always summable to the same value
by the method of the first arithmetic mean*.

To prove the last statement, suppose that $\lim_{n \to \infty} s_n$
$= A$, and let ϵ be an arbitrary positive quantity. There
is a number m such that $\left| s_n - A \right| < \frac{1}{2}\epsilon$ for all $n \geqq m$. The
difference between σ_n and A can be written in the form

$$\sigma_n - A = (s_0 + s_1 + \cdots + s_{n-1} - nA)/n$$

$$= \frac{1}{n} \sum_{k=0}^{n-1} (s_k - A).$$

When $n > m$ this is the same as

$$\frac{1}{n} \sum_{k=0}^{m-1} (s_k - A) + \frac{1}{n} \sum_{k=m}^{n-1} (s_k - A).$$

In each term of the second sum $|s_k - A| < \frac{1}{2}\epsilon$, the number of terms does not exceed n, and

$$\frac{1}{n} \sum_{k=m}^{n-1} |s_k - A| < \frac{1}{2}\epsilon$$

for all $n > m$. The first sum is independent of n, the result of dividing it by n approaches zero when n becomes infinite, and

$$\left| \frac{1}{n} \sum_{k=0}^{m-1} (s_k - A) \right| < \frac{1}{2}\epsilon$$

if n is sufficiently large. So $|\sigma_n - A| < \epsilon$ for all sufficiently large values of n, which means that $\lim_{n \to \infty} \sigma_n = A$.

If each term of the series is a bounded function of a variable x, the quantities s_k and σ_n and A also being then functions of x, *and if the series is uniformly convergent*, m can be taken independent of x, and the above proof then shows that *the approach of σ_n to its limit is uniform*.

18. Fejér's theorem for a continuous function. By a notable theorem due to Fejér,* *the Fourier series for an arbitrary continuous function f(x) of period 2π is uniformly summable by the method of the first arithmetic mean to the value f(x)*. This is a part of the content of a more general formulation which will not be discussed here. The theorem stated, like that of Weierstrass in

---

* L. Fejér, *Untersuchungen über Fouriersche Reihen*, Mathematische Annalen, vol. 58 (1904), pp. 51–69.

§ 14, draws its significance from the fact that the Fourier series itself may not be convergent.

Let $s_n(x)$ be the partial sum of the series as usual, and let

$$\sigma_n(x) = (1/n)\left[s_0(x) + s_1(x) + \cdots + s_{n-1}(x)\right].$$

By the use of (20) for the representation of $s_k(x)$,

$$\sigma_n(x) = \frac{1}{n\pi} \int_{-\pi}^{\pi} \frac{f(x+u)}{2\sin\frac{1}{2}u} \sum_{k=0}^{n-1} \sin\left(k+\tfrac{1}{2}\right)u \, du.$$

Let the sum of sines in the integrand be multiplied by $2\sin\frac{1}{2}u$, and let the identity

$$2\sin\tfrac{1}{2}u \sin\left(k+\tfrac{1}{2}\right)u = \cos ku - \cos\left(k+1\right)u$$

be used in evaluating the result. It is found that

$$(31) \quad 2\sin\tfrac{1}{2}u\sum_{k=0}^{n-1}\sin\left(k+\tfrac{1}{2}\right)u = 1 - \cos nu = 2\sin^2\left(\tfrac{1}{2}nu\right).$$

Consequently

$$(32) \qquad \sigma_n(x) = \frac{1}{n\pi}\int_{-\pi}^{\pi} f(x+u)\,\frac{\sin^2\frac{1}{2}nu}{2\sin^2\frac{1}{2}u}\,du.$$

From (18),

$$\sin\left(k+\tfrac{1}{2}\right)u = 2\sin\tfrac{1}{2}u\left[\tfrac{1}{2}+\cos u+\cos 2u+ \cdots +\cos ku\right],$$

the expression in brackets being understood to reduce to the single term $\frac{1}{2}$ for $k=0$. By writing this relation for $k = 0, 1, \cdots, n-1$ successively and adding it is seen that

$$(33) \quad \begin{aligned} &\sum_{k=0}^{n-1}\sin\left(k+\tfrac{1}{2}\right)u \\ &= 2\sin\tfrac{1}{2}u\left[\frac{n}{2}+(n-1)\cos u+ \cdots + \cos(n-1)u\right]. \end{aligned}$$

By combination of (31) and (33),

$$\frac{\sin^2 \frac{1}{2}nu}{2 \sin^2 \frac{1}{2}u} = \frac{n}{2} + (n-1) \cos u + \cdots + \cos (n-1)u, .$$

whence

$$(34) \quad \frac{1}{n\pi} \int_{-\pi}^{\pi} \frac{\sin^2 \frac{1}{2}nu}{2 \sin^2 \frac{1}{2}u} \, du = \frac{1}{n\pi} \int_{-\pi}^{\pi} \frac{n}{2} \, du = 1,$$

a fact which can also be deduced immediately from (32) by noting that each partial sum of the Fourier series for a constant reduces identically to that constant, and the same is true therefore of the arithmetic mean of any number of such partial sums.

An important property of $\sigma_n(x)$, on which the generality of its convergence depends, is a consequence of the fact that the trigonometric fraction in the integrand of (32), unlike the corresponding factor in (20), is everywhere non-negative. If M is the maximum of $|f(x)|$ (or an upper bound for $|f(x)|$, if $f(x)$ is bounded without being continuous)

$$|\sigma_n(x)| \leq \frac{M}{n\pi} \int_{-\pi}^{\pi} \frac{\sin^2 \frac{1}{2}nu}{2 \sin^2 \frac{1}{2}u} \, du,$$

which means in consequence of (34) that

$$(35) \qquad\qquad |\sigma_n(x)| \leq M$$

for all values of n and x. A corresponding calculation based on (20) would be complicated by the necessity of replacing the fraction by its absolute value (see §20 below).

Let it be assumed now that $f(x)$ is of period 2π and continuous. Let ϵ be an arbitrary positive quantity. Let $g(x)$ be a broken-line function constructed as in §14 to approximate $f(x)$, but with $\frac{1}{3}\epsilon$ instead of $\frac{1}{2}\epsilon$ as upper bound for the difference, so that

$$\left| f(x) - g(x) \right| < \tfrac{1}{3}\epsilon.$$

The quantity $f(x) - g(x)$ being denoted by $h(x)$, let arithmetic means corresponding to $\sigma_n(x)$ be formed for the functions $g(x)$ and $h(x)$, and let these means be denoted by $\sigma_{n1}(x)$ and $\sigma_{n2}(x)$ respectively. Then $f(x) = g(x) + h(x)$, $\sigma_n(x) = \sigma_{n1}(x) + \sigma_{n2}(x)$, and

$$f(x) - \sigma_n(x) = \left[g(x) - \sigma_{n1}(x) \right] + \left[h(x) - \sigma_{n2}(x) \right].$$

By (35), as applied to $h(x)$ and $\sigma_{n2}(x)$,

$$\left| \sigma_{n2}(x) \right| < \tfrac{1}{3}\epsilon, \qquad \left| h(x) - \sigma_{n2}(x) \right| < \tfrac{2}{3}\epsilon,$$

for all n and all x. By §11 the Fourier series for $g(x)$ is uniformly convergent to the value $g(x)$; by §17, therefore, $\sigma_{n1}(x)$ approaches $g(x)$ uniformly, and

$$\left| g(x) - \sigma_{n1}(x) \right| < \tfrac{1}{3}\epsilon$$

everywhere, if n is large enough. Consequently, for n sufficiently large,

$$\left| f(x) - \sigma_n(x) \right| < \epsilon$$

for all values of x.

19. Proof of Weierstrass's theorem by means of de la Vallée Poussin's integral. An alternative proof of Weierstrass's theorem which is of interest in itself, and can also be made to give additional information about the convergence of Fourier series, is based on an integral formula due to de la Vallée Poussin.[*]

---

[*] C. de la Vallée Poussin, *Sur l'approximation des fonctions d'une variable réelle et de leurs dérivées par des polynômes et des suites limitées de Fourier*, Bulletins de l'Académie Royale de Belgique, Classe des Sciences, 1908, pp. 193–254; pp. 227–238.

Let $f(x)$ *be a continuous function of period* 2π. The formula is

(36) $$V_n(x) = \frac{1}{H_n} \int_{-\pi/2}^{\pi/2} f(x + 2u) \cos^{2n} u \, du,$$

in which H_n is a constant defined by

(37) $$H_n = \int_{-\pi/2}^{\pi/2} \cos^{2n} u \, du = 2 \int_{0}^{\pi/2} \cos^{2n} u \, du.$$

The function $V_n(x)$ *is a trigonometric sum in x, of the nth order at most.* This may be seen as follows. Let the variable of integration in (36) be changed by the substitution $x + 2u = t$. In the new integrand, $f(t)$ has the period 2π. The same is true of $\cos^{2n} \frac{1}{2}(t-x)$; increase of t by 2π reverses the algebraic sign of $\cos \frac{1}{2}(t-x)$, but leaves any even power of it unchanged. So the interval of integration with respect to t can be taken as $(-\pi, \pi)$ instead of $(x-\pi, x+\pi)$:

$$V_n(x) = \frac{1}{2H_n} \int_{-\pi}^{\pi} f(t) \cos^{2n}\left(\frac{t-x}{2}\right) dt.$$

Since $\cos^2 \frac{1}{2}u = \frac{1}{2}(1 + \cos u)$, the function $\cos^{2n} \frac{1}{2}u$ can be written in the form $2^{-n}(1 + \cos u)^n$. This is a cosine sum of the nth order in u (the phrase *cosine sum* being used for brevity to mean *trigonometric sum involving only cosines*). For it follows from the identity

$$\cos px \cos qx = \tfrac{1}{2}\cos (p+q)x + \tfrac{1}{2}\cos (p-q)x$$

that the product of a cosine sum of order n_1 and a cosine sum of order n_2 is a cosine sum of order $n_1 + n_2$, and hence by induction that the successive powers of $1 + \cos u$ are cosine sums of corresponding order. So $\cos^{2n} \frac{1}{2}(t-x)$ is a cosine sum of the nth order in $t-x$. By means of the representation

$$\cos k(t - x) = \cos kt \cos kx + \sin kt \sin kx$$

for each value of k the cosine sum in $t - x$ can be written as a trigonometric sum of the nth order in x, with coefficients which are functions of t. The last description applies also to the product $f(t) \cos^{2n} \frac{1}{2}(t-x)$. So $V_n(x)$ is a trigonometric sum in x of the nth order at most, with constant coefficients

This fact being established, further discussion will be based on the original formula (36). Multiplication of (37) by $f(x)/H_n$, which is independent of the variable of integration, gives

$$f(x) = \frac{1}{H_n} \int_{-\pi/2}^{\pi/2} f(x) \cos^{2n} u \, du.$$

By subtraction of this from (36),

$$V_n(x) - f(x) = \frac{1}{H_n} \int_{-\pi/2}^{\pi/2} \left[f(x + 2u) - f(x) \right] \cos^{2n} u \, du.$$

Let $f(x)$ be subjected to the additional restriction that

(38) $$\left| f(x_2) - f(x_1) \right| \leqq \lambda \left| x_2 - x_1 \right|$$

for all x_1 and x_2, λ being a constant. Then

$$\left| f(x + 2u) - f(x) \right| \leqq 2\lambda \left| u \right|,$$

$$\left| V_n(x) - f(x) \right| \leqq \frac{2\lambda}{H_n} \int_{-\pi/2}^{\pi/2} \left| u \right| \cos^{2n} u \, du$$

$$= \frac{4\lambda}{H_n} \int_0^{\pi/2} u \cos^{2n} u \, du.$$

Since

$$\frac{d}{du} \left(\frac{u}{\sin u} \right) = \frac{\sin u - u \cos u}{\sin^2 u} = \frac{\cos u}{\sin^2 u} (\tan u - u) > 0$$

for $0 < u < \pi/2$, the function $u/\sin u$ increases throughout this interval, and

(39) $$\frac{u}{\sin u} \leqq \frac{\pi/2}{\sin (\pi/2)} = \frac{\pi}{2}, \qquad u \leqq \frac{\pi}{2} \sin u.$$

(The derivative is obviously positive also for $\pi/2 \leqq u < \pi$, since $\sin u$ is positive and $(-u \cos u)$ is non-negative there.) So

(40) $$\left| V_n(x) - f(x) \right| \leqq \frac{2\lambda\pi}{H_n} \int_0^{\pi/2} \sin u \cos^{2n} u \, du.$$

Let the last integral be denoted by h_n. It can of course be evaluated explicitly at a single stroke, but the requisite comparison of its magnitude with that of H_n can be obtained more readily by a different procedure.

If $g_1(x)$ and $g_2(x)$ are any two functions such that g_1^2 and g_2^2 as well as g_1 and g_2 are integrable over an interval (a, b), and if μ is a parameter, the integral

$$\int_a^b [g_1(x) - \mu g_2(x)]^2 dx$$

is non-negative for all real values of μ. So the quadratic equation in μ obtained by setting the integral equal to 0 can not have distinct real roots, and its discriminant must be negative or zero, i.e.

$$\left[\int_a^b g_1(x) g_2(x) dx \right]^2 \leqq \int_a^b [g_1(x)]^2 dx \int_a^b [g_2(x)]^2 dx.$$

This general relation is known as *Schwarz's inequality*.*

* The name is applied also to other relations of essentially the same form, for example those obtained by application of similar reasoning to double or multiple integrals.

It will be useful here, and on more than one occasion in a later chapter.

Let the integrand in h_n be regarded as the product of the factors $\sin u \cos^n u$ and $\cos^n u$. By Schwarz's inequality,

$$h_n^2 \leqq \int_0^{\pi/2} \sin^2 u \cos^{2n} u \, du \int_0^{\pi/2} \cos^{2n} u \, du.$$

The second integral on the right is equal to $\frac{1}{2}H_n$. Let k_n denote the other. Integration by parts with factors $\sin u$ and $\sin u \cos^{2n} u \, du$ gives

$$k_n = \frac{1}{2n+1} \left[-\sin u \cos^{2n+1} u \right]_0^{\pi/2} + \frac{1}{2n+1} \int_0^{\pi/2} \cos^{2n+2} u \, du$$

$$= \frac{1}{2} H_{n+1}/(2n+1) < H_{n+1}/(4n) < H_n/(4n),$$

the last inequality resulting from the fact that $\cos^{2n+2}u < \cos^{2n} u$ throughout the interior of the interval of integration. So

$$h_n^2 < H_n^2/(8n), \qquad h_n < H_n/(2^{3/2} n^{1/2}) < \frac{1}{2} H_n/n^{1/2}.$$

By combination of this with (40), since h_n denotes the value of the integral in (40),

$$(41) \qquad \left| V_n(x) - f(x) \right| \leqq \lambda \pi/n^{1/2}.$$

From (41) it follows that $V_n(x)$ approaches $f(x)$ uniformly as n becomes infinite. This to be sure is under the hypothesis that $f(x)$ satisfies the supplementary condition (38). But if $f(x)$ is an arbitrary continuous function of period 2π, and if a broken-line function $g(x)$ is constructed as in § 14 to differ from $f(x)$ by less than $\frac{1}{2}\epsilon$, this broken-line function satisfies a condition of the form (38); the trigonometric sum defined by substituting $g(x+2u)$ for $f(x+2u)$ in (36) differs from $g(x)$ by less than $\frac{1}{2}\epsilon$ when n is sufficiently large, and then differs from $f(x)$ by less than ϵ. *Thus another proof of Weierstrass's theorem is obtained.*

It would have been almost as easy to give the proof by showing that the sum $V_n(x)$ defined by (36) for the original $f(x)$ converges uniformly toward $f(x)$, without any hypothesis on $f(x)$ beyond those of continuity and periodicity. The purpose of this section however is not merely to demonstrate again a theorem which has already been proved, but to derive for use in §21 the specific inequality given by (41) for the order of magnitude of the error, in the case of a function $f(x)$ which satisfies (38).

20. The Lebesgue constants. Let $f(x)$ be any function which is bounded and integrable over $(-\pi, \pi)$. Let M be an upper bound for its absolute value, i. e. a constant such that $\left| f(x) \right| \leqq M$ for $-\pi \leqq x \leqq \pi$. Then, from (20),

$$(42) \qquad \left| s_n(x) \right| \leqq \lambda_n M$$

throughout the interval, if

$$(43) \qquad \begin{aligned} \lambda_n &= \frac{1}{\pi} \int_{-\pi}^{\pi} \left| \frac{\sin (n + \frac{1}{2})u}{2 \sin \frac{1}{2}u} \right| du \\ &= \frac{2}{\pi} \int_{0}^{\pi} \left| \frac{\sin (n + \frac{1}{2})u}{2 \sin \frac{1}{2}u} \right| du. \end{aligned}$$

The quantities λ_n (which are independent of $f(x)$) are called the *Lebesgue constants* of the Fourier series. They do not remain finite as n becomes infinite, and (42) is therefore a materially less simple relation than the inequality (35) for $\sigma_n(x)$. The fact that $\lim_{n \to \infty} \lambda_n = \infty$ will not be proved here. It will be shown on the other hand that λ_n *does not become infinite faster than a quantity of the order of magnitude* of* $\log n$.

Let the integral from 0 to π in (43) be regarded as the sum of integrals extended from 0 to $1/n$ and from $1/n$ to π. By (18), since $\left| \cos ku \right| \leqq 1$ for each k,

$$\left| \frac{\sin (n + \frac{1}{2})u}{2 \sin \frac{1}{2}u} \right| \leqq n + \tfrac{1}{2}$$

for all values of u. Application of this inequality in the interval $(0, 1/n)$ gives

$$\int_0^{1/n} \left| \frac{\sin (n + \frac{1}{2})u}{2 \sin \frac{1}{2}u} \right| du \leqq \frac{1}{n} (n + \tfrac{1}{2}) = 1 + \frac{1}{2n} \cdot$$

In the other interval $\left| \sin (n+\frac{1}{2})u \right| \leqq 1$, and by (39), with u replaced by $\frac{1}{2}u$ (which does not exceed $\frac{1}{2}\pi$),

$$\sin \tfrac{1}{2}u \geqq u/\pi, \qquad 1/\sin \tfrac{1}{2}u \leqq \pi/u;$$

the last relations of course become equalities for $u = \pi$. So

$$\int_{1/n}^{\pi} \left| \frac{\sin (n + \frac{1}{2})u}{2 \sin \frac{1}{2}u} \right| du \leqq \frac{\pi}{2} \int_{1/n}^{\pi} \frac{du}{u} = \frac{\pi}{2} (\log \pi + \log n).$$

Hence

$$\lambda_n \leqq \log n + \log \pi + \frac{2}{\pi} \left(1 + \frac{1}{2n} \right).$$

When $n \geqq 2$, $1 \leqq \log n / \log 2$, and

$$\log \pi + \frac{2}{\pi} \left(1 + \frac{1}{2n} \right) \leqq \log \pi + \frac{2}{\pi} \cdot \frac{5}{4}$$

$$\leqq \left(\log \pi + \frac{5}{2\pi} \right) \frac{\log n}{\log 2} ;$$

---

* See H. Lebesgue, *Sur les intégrales singulières*, Annales de la Faculté de Toulouse, (3), vol. 1 (1909), pp. 25–117; pp. 116–117.

if the constant $\{\log \pi + [5/(2\pi)]\}/\log 2$ in the last member is denoted by $C-1$,

$$\lambda_n \leq C \log n$$

for $n \geq 2$.

It follows further, since $|s_n(x)| \leq \lambda_n M$ and $|f(x)| \leq M$, that when $n \geq 2$

$$|f(x) - s_n(x)| \leq M(1 + C \log n) \leq M(\log n/\log 2 + C \log n),$$

or, if C' denotes the constant $(1/\log 2) + C$,

$$(44) \qquad |f(x) - s_n(x)| \leq C'M \log n.$$

21. Proof of uniform convergence by the method of Lebesgue. If $T_n(x)$ is a trigonometric sum of the nth order,*

$$T_n(x) = \frac{\alpha_0}{2} + \sum_{k=1}^{n} (\alpha_k \cos kx + \beta_k \sin kx),$$

and if this function is expanded in a Fourier series, it is found at once that the coefficients a_k, b_k in the expansion are the same as α_k, β_k respectively for $k \leq n$, and are zero for $k > n$. *A trigonometric sum is its own Fourier series.* The corresponding partial sum $s_n(x)$ is identical with $T_n(x)$.

Let $f(x)$ be a function of period 2π, for simplicity continuous, and suppose that a trigonometric sum $T_n(x)$ of the nth order and a number ϵ_n are such that

$$(45) \qquad |f(x) - T_n(x)| \leq \epsilon_n$$

for all x. Let $f(x) - T_n(x) = r_n(x)$, and let $s_{n1}(x)$ be the partial sum of the nth order in the Fourier series for

---

* In this section a "trigonometric sum of the nth order" will be understood to be one of the nth order *at most*; the possibility that $\alpha_n = \beta_n = 0$ is not excluded.

$r_n(x)$, while $s_n(x)$ is now the partial sum of the series for $f(x)$; as noted above, the corresponding partial sum in the case of $T_n(x)$ is $T_n(x)$ itself. Then, as $f(x) = r_n(x) + T_n(x)$, and the partial sum of the series for a sum of two functions is obtained by adding the partial sums of the series formed for the two functions separately,

$$s_n(x) = s_{n1}(x) + T_n(x),$$

and

$$f(x) - s_n(x) = r_n(x) - s_{n1}(x).$$

Thus the error of $s_n(x)$ as an approximation to $f(x)$ is the same as that of $s_{n1}(x)$ with respect to $r_n(x)$, in consequence of the fact that the corresponding approximation to $T_n(x)$ involves no error at all; and $r_n(x)$ may be a function which is small everywhere. This simple observation, taken with the results of the preceding section, as was pointed out by Lebesgue,[*] leads to an effective method of proving convergence.

It is to be noted that the constant C' in (44) is merely a number which can be calculated once for all, being not only independent of x and n, but independent also of the function $f(x)$. By application of (44), with reference to (45),

$$\left| r_n(x) - s_{n1}(x) \right| \leqq C' \epsilon_n \log n,$$

and the quantity on the right is an upper bound for $\left| f(x) - s_n(x) \right|$ at the same time:

If $f(x)$ can be uniformly approximated by a trigonometric sum of the nth order with an error not exceeding ϵ_n, and if $s_n(x)$ is the partial sum of the Fourier series for $f(x)$,

[*] See second preceding footnote.

$$\left| f(x) - s_n(x) \right| \leqq C' \epsilon_n \log n$$

for all values of x.

Let it be supposed now that $f(x)$ satisfies (38). Then if the trigonometric sum $V_n(x)$ of §19 is taken as $T_n(x)$, the relation (41) means that (45) is satisfied with $\epsilon_n = \lambda \pi / n^{1/2}$. In this case

$$(46) \qquad \left| f(x) - s_n(x) \right| \leqq C' \lambda \pi \log n / n^{1/2}$$

for all x, and $\log n / n^{1/2}$ approaches zero as n becomes infinite. The particular form that has been obtained for the right-hand member is not ultimately significant; it can be shown by other methods that the series is in fact more rapidly convergent* than is apparent from (46). The reasoning as given, however, constitutes a proof that *if $f(x)$ is a function of period 2π such that*

$$\left| f(x_2) - f(x_1) \right| \leqq \lambda \left| x_2 - x_1 \right|$$

for all x_1 and x_2, λ being a constant, its Fourier series converges uniformly to the value $f(x)$.

This hypothesis is considerably more general than those under which uniform convergence was proved in §11.

SUPPLEMENTARY REFERENCES (see Bibliography at end of book): Zygmund; Carslaw; Lebesgue; Tonelli; Bôcher; Byerly; Churchill; Kellogg, pp. 355–359; Titchmarsh; Whittaker and Watson, Chapter IX; Hobson (1); Fejér; Courant-Hilbert; Riemann-Weber; Pólya-Szegö; Kaczmarz-Steinhaus.

* See H. Lebesgue, *Sur la représentation trigonométrique approchée des fonctions satisfaisant à une condition de Lipschitz*, Bulletin de la Société Mathématique de France, vol. 38 (1910), pp. 184–210; pp. 199–202; D. Jackson, *The Theory of Approximation*, American Mathematical Society Colloquium Publications, vol. 11, New York, 1930, pp. 18–23.

CHAPTER II

LEGENDRE POLYNOMIALS

1. Preliminary orientation. A type of series resembling Fourier series in many respects is one in which the sines and cosines are replaced by certain polynomials, called Legendre polynomials. These can be defined in a variety of ways, and their principal properties are so interrelated that the arrangement of them in logical sequence admits a multiplicity of permutations. The order of exposition followed here for a rapid derivation of the principal facts is chosen rather for its convenience from the point of view of the results to be obtained than for any obvious motivation of its steps in advance. One of the possible alternative arrangements will be found in the later chapters on orthogonal polynomials in general and Jacobi polynomials in particular, which include the Legendre polynomials as a special case.

2. Definition of the Legendre polynomials by means of the generating function. Let

$$(1) \qquad H(x, r) = \frac{1}{(1 - 2xr + r^2)^{1/2}}.$$

As a function of r, this can be developed in a power series for sufficiently small values of the variable. Let the coefficients, which (except for the constant term) will depend on x, be denoted by $P_n(x)$, $n = 0, 1, 2, \cdots$, so that

$$(2) \qquad H(x, r) = P_0(x) + P_1(x)r + P_2(x)r^2 + \cdots .$$

They can be calculated formally by setting $y = 2xr - r^2$

and substituting the expansions of the successive powers of $2xr - r^2$ in the binomial series for $(1 - y)^{-1/2}$. The first few coefficients are found to be

$$P_0(x) = 1, \qquad P_1(x) = x, \qquad P_2(x) = \tfrac{1}{2}(3x^2 - 1),$$
$$P_3(x) = \tfrac{1}{2}(5x^3 - 3x), \qquad P_4(x) = \tfrac{1}{8}(35x^4 - 30x^2 + 3).$$

Certain general facts with regard to the functions $P_n(x)$ are immediately apparent. Each is a polynomial in x. Furthermore, $P_n(x)$ is of the nth degree. For the exponent of x never exceeds that of r in any term of the expansion of any power of y, and as x^n is associated with r^n only in the initial term of the expansion of y^n this term can not be cancelled by terms from any other source. *The polynomials $P_0(x)$, $P_1(x)$ $P_2(x)$, $\cdots$, each of degree indicated by its subscript, are the Legendre polynomials.* Since each term is of even degree in x and r together, even powers of x are associated only with even powers of r, and odd powers of x with odd powers of r; each polynomial $P_n(x)$ contains terms of even or of odd degree exclusively according as n is even or odd, and

$$(3) \qquad P_n(- x) \equiv (- 1)^n P_n(x).$$

For $x = 1$, $H(x, r)$ reduces to

$$(1 - r)^{-1} = 1 + r + r^2 + \cdots,$$

so that $P_n(1) = 1$ for each value of n. From (3) it follows then that $P_n(-1) = (-1)^n$.

The function $H(x, r)$ is called a *generating function* for the Legendre polynomials.

3. Recurrence formula. By explicit differentiation of $H(x, r)$ it is seen that

$$(4) \qquad (1 - 2xr + r^2) \frac{\partial H}{\partial r} - (x - r)H = 0.$$

If the left-hand member of (4) is written as a power series in r by substitution of the representation (2) for H and the differentiated series

$$P_1(x) + 2P_2(x)r + 3P_3(x)r^2 + \cdots$$

for $\partial H/\partial r$, the coefficient of r^n must be zero for each n and for all values of x. The vanishing of this coefficient is expressed by the identity

$$(5) \quad (n + 1)P_{n+1}(x) - (2n + 1)xP_n(x) + nP_{n-1}(x) = 0.$$

This is a *relation of recurrence* connecting any three successive Legendre polynomials. It can be used for the explicit calculation of $P_2(x)$, $P_3(x)$, $\cdots$ when the first two polynomials $P_0(x) = 1$, $P_1(x) = x$ are known, and will be further applied in a variety of ways. Combined with the initial determination of P_0 and P_1, it serves incidentally to verify the observations of the preceding section that the exponents of the powers of x appearing in $P_n(x)$ are all even or all odd and that $P_n(1) = 1$.

Let A_n denote the leading coefficient, i. e. the coefficient of x^n, in $P_n(x)$ for each value of n. Then the coefficient of x^{n+1} in the left member of (5) is $(n+1)A_{n+1} - (2n+1)A_n$. Since this must vanish,

$$(6) \qquad A_{n+1} = \frac{2n + 1}{n + 1} A_n.$$

With the value $A_0 = 1$ for a beginning, successive substitutions in this formula give $A_1 = 1$, $A_2 = 3/2$, $A_3 = 5/2$, and in general, by induction,

$$
(7) \quad
\begin{aligned}
A_n &= 1 \cdot \frac{3}{2} \cdot \frac{5}{3} \cdot \frac{7}{4} \cdots \frac{(2n - 1)}{n} \\
&= \frac{1 \cdot 3 \cdot 5 \cdots (2n - 1)}{n!}.
\end{aligned}
$$

(The relation (5) holds for $n = 0$ as well as for positive n, if $0 \cdot P_{-1}(x)$ is interpreted as having the value 0, and the derivation of (6) from (5) can therefore be regarded as valid for $n = 0$ likewise.)

4. Differential equation and related formulas. By differentiation of (1) with respect to x,

$$(8) \qquad (1 - 2xr + r^2) \frac{\partial H}{\partial x} - rH = 0.$$

Comparison of (4) with (8) (or of the explicit formulas for $\partial H / \partial r$ and $\partial H / \partial x$) gives the simpler relation

$$(9) \qquad r \frac{\partial H}{\partial} - (x - r) \frac{\partial H}{\partial x} = 0.$$

By equating to zero the coefficient of r^n in the power series for the left-hand member of (9) it is found that

$$(10) \qquad nP_n(x) - xP_n'(x) + P_{n-1}'(x) = 0.$$

Still another partial differential equation satisfied by H is

$$(11) \qquad r \frac{\partial}{\partial r} (rH) - (1 - rx) \frac{\partial H}{\partial x} = 0;$$

this is obtained by writing $r(\partial/\partial r)(rH) = r(r\partial H/\partial r + H)$ and substituting the alternative expressions for $r\partial H/\partial r$ and rH given by (9) and (8) respectively. Representation by power series in (11) gives for the vanishing of the coefficient of r^n

$$(12) \qquad nP_{n-1}(x) - P_n'(x) + xP_{n-1}'(x) = 0,$$

and for the coefficient of r^{n+1} (obtained by substitution of $n+1$ for n in the preceding)

(13) $\qquad (n + 1)P_n(x) - P'_{n+1}(x) + xP'_n(x) = 0.$

These identities are of secondary interest in themselves; combination of them leads to others which are of greater importance.

By addition of (10) and (13) and transposition of terms,

(14) $\qquad P'_{n+1}(x) - P'_{n-1}(x) = (2n + 1)P_n(x).$

A still more important relation is obtained by eliminating $P_{n-1}(x)$ and $P'_{n-1}(x)$ between (10) and (12). This can be accomplished by substituting in (12) the value of $P'_{n-1}(x)$ given by (10), differentiating, and substituting from (10) again for the $P'_{n-1}(x)$ introduced by the differentiation. An alternative formulation of the procedure consists in representing the left-hand members of (10) and (12) by $\Phi_1(x)$ and $\Phi_2(x)$ respectively, and equating to zero the expression

$$\frac{d}{dx}(x\Phi_1 - \Phi_2) + n\Phi_1.$$

The result is

(15) $\quad (1 - x^2)P''_n(x) - 2xP'_n(x) + n(n + 1)P_n(x) = 0.$

The Legendre polynomial $P_n(x)$ satisfies this linear homogeneous differential equation.

From the point of view of demonstration it is not necessary for the justification of the above steps to refer to theorems on the differentiation of infinite series as such; the reasoning can be based on the definition of the various coefficients by the values of the appropriate derivatives for $r = 0$, in accordance with the usual rule for setting up a Maclaurin series.

5. Orthogonality. Let $y_m = P_m(x)$, $y_n = P_n(x)$. Let the differential equations (15) satisfied by y_m and y_n be multiplied by y_n and y_m respectively:

$$(16) \quad \begin{aligned} y_n[(1 - x^2)y_m'' - 2xy_m' + m(m + 1)y_m] &= 0, \\ y_m[(1 - x^2)y_n'' - 2xy_n' + n(n + 1)y_n] &= 0. \end{aligned}$$

Let $w = y_m' y_n - y_n' y_m$; then $dw/dx = y_m'' y_n - y_n'' y_m$, and the result of subtracting one of the equations (16) from the other can be written in the form

$$\begin{aligned} (1 - x^2)w' - 2xw &= \frac{d}{dx}\left[(1 - x^2)w\right] \\ &= [n(n + 1) - m(m + 1)]y_m y_n. \end{aligned}$$

Consequently, by integration from -1 to 1,

$$\left[(1 - x^2)w\right]_{-1}^{1} = [n(n + 1) - m(m + 1)]\int_{-1}^{1} y_m y_n dx.$$

The expression in brackets on the left vanishes for $x = \pm 1$. The bracket on the right is not zero if the non-negative integers m and n are distinct from each other; if $n > m$, then $n(n+1) > m(m+1)$, and vice versa. Therefore

$$(17) \qquad \int_{-1}^{1} P_m(x)P_n(x)dx = 0$$

if $m \neq n$. *Any two Legendre polynomials of different degrees are orthogonal to each other over the interval* $(-1, 1)$.

Since $P_n(x)$ contains x^n with a non-vanishing coefficient for each value of n, the identities expressing the Legendre polynomials in terms of powers of x can be successively solved for the latter. Each power of x can be expressed in terms of Legendre polynomials, and any polynomial of the nth degree can be expressed as a linear combination of $P_0(x)$, $P_1(x)$, $\cdots$, $P_n(x)$ with constant coefficients. It follows from (17) therefore that $P_n(x)$ *is orthogonal over the interval* $(-1, 1)$ *to every polynomial of lower degree*; if $q(x)$ is any such polynomial,

$$\int_{-1}^{1} P_n(x)q(x)dx = 0.$$

Explicitly, the representations of the first few powers of x in terms of the P's are found to be

$$1 = P_0(x), \qquad x = P_1(x), \qquad x^2 = \tfrac{2}{3}P_2(x) + \tfrac{1}{3}P_0(x),$$
$$x^3 = \tfrac{2}{5}P_3(x) + \tfrac{3}{5}P_1(x),$$
$$x^4 = \tfrac{1}{35}\left[8P_4(x) + 20P_2(x) + 7P_0(x)\right].$$

6. Normalizing factor. When $m = n$ the integral in (17) is certainly not zero, since the integrand is nonnegative and in general different from zero. Let

$$C_n = \int_{-1}^{1} \left[P_n(x)\right]^2 dx.$$

The leading coefficients in $P_n(x)$ and $P_{n-1}(x)$ being denoted by A_n and A_{n-1} respectively, in accordance with the notation of §3, the combination

$$P_n(x) - (A_n/A_{n-1})xP_{n-1}(x)$$

contains no term in x^n, and is a polynomial $q(x)$ of lower degree, òrthogonal to $P_n(x)$. Then

$$P_n(x) = (A_n/A_{n-1})xP_{n-1}(x) + q(x),$$
$$C_n = \int_{-1}^{1} P_n(x)\left[(A_n/A_{n-1})xP_{n-1}(x) + q(x)\right]dx$$
$$= \frac{A_n}{A_{n-1}}\int_{-1}^{1} xP_n(x)P_{n-1}(x)dx.$$

By the recurrence formula (5),

$$xP_n(x) = \frac{n+1}{2n+1} P_{n+1}(x) + \frac{n}{2n+1} P_{n-1}(x);$$

and $P_{n+1}(x)$ is orthogonal to $P_{n-1}(x)$. Hence

$$C_n = \frac{n}{2n+1} \cdot \frac{A_n}{A_{n-1}} \int_{-1}^{1} [P_{n-1}(x)]^2 dx = \frac{n}{2n+1} \cdot \frac{A_n}{A_{n-1}} \cdot C_{n-1}.$$

By (6), with n replaced by $n-1$, $A_n = (2n-1)A_{n-1}/n$, which makes

$$C_n = \frac{2n-1}{2n+1} C_{n-1}.$$

Since $C_0 = 2$, by explicit evaluation of the integral, it follows that $C_1 = 2/3$, $C_2 = 2/5$, and by induction

$$(18) \qquad C_n = \int_{-1}^{1} [P_n(x)]^2 dx = \frac{2}{2n+1}.$$

The polynomial $p_n(x) = [(2n+1)/2]^{1/2} P_n(x)$, in which the constant factor is determined so that

$$\int_{-1}^{1} [p_n(x)]^2 dx = 1,$$

is the *normalized Legendre polynomial* of the nth degree. The introduction of the normalized polynomials is convenient for some purposes as a matter of notation. The discussion will be carried on for the present, however, in terms of the polynomials $P_n(x)$ as originally defined.

It is an almost immediate consequence of the property of orthogonality that the roots of the equation $P_n(x) = 0$ are all real, all distinct, and all interior to the interval $(-1, 1)$. The proof is omitted here, to be given later under conditions of more general applicability (see Chapter VII, §8).

7. Expansion of an arbitrary function in series. On the basis of the facts now available an arbitrary function defined on the interval $(-1, 1)$ can be formally expanded in a series of Legendre polynomials. The resemblance to Fourier series is already apparent in the manner of determination of the coefficients. If the expansion is

$$(19) \quad f(x) = a_0 P_0(x) + a_1 P_1(x) + a_2 P_2(x) + \cdots ,$$

let this be multiplied by $P_n(x)$ and integrated term by term from -1 to 1, on the assumption that the series is such that the indicated process is legitimate. Each integral on the right is zero, because of the orthogonality of the P's, except the one containing P_k^2, and when this integral is evaluated by means of (18) the whole right-hand member reduces to $2a_k/(2k+1)$. So the general coefficient in the series is given by

$$(20) \qquad a_k = \frac{2k+1}{2} \int_{-1}^{1} f(x) P_k(x) dx.$$

As in the case of Fourier series, a set of coefficients can be defined for any function which is integrable over the interval, without any further assumption *a priori*, and the series thus obtained can then be examined with regard to its convergence and its validity as a representation of the function.

If $f(x)$ is itself a polynomial $\pi_n(x)$ of the nth (or lower) degree, it is known already that a representation of the form (19) exists with a finite sum instead of an infinite series on the right-hand side. The derivation of (20) amounts in this case to a definitive proof of that formula, as far as values of $k \leqq n$ are concerned, without question of convergence, while the formula automatically gives $a_k = 0$ for $k > n$. The coefficients in the last lines of §5 can be checked by explicit substitution in (20) of $f(x) \equiv 1, x, \cdots, x^4$.

8. Christoffel's identity. Let $s_n(x)$ denote the partial sum of the series for an arbitrary $f(x)$ through terms of the nth degree:

$$(21) \quad s_n(x) = a_0 P_0(x) + a_1 P_1(x) + \cdots + a_n P_n(x).$$

If (20) is written with t instead of x as variable of integration, the formulas for the coefficients can be substituted explicitly in (21) to give

$$(22) \qquad s_n(x) = \int_{-1}^{1} f(t) K_n(x, t) dt,$$

with

$$K_n(x, t) = K_n(t, x) = \sum_{k=0}^{n} \frac{2k+1}{2} P_k(t) P_k(x).$$

The derivation of a formula corresponding to (19) of Chapter I depends on an identity for the evaluation of $K_n(x, t)$.

From the recurrence formula (5), with replacement of n by k, transposition of terms, and multiplication by $P_k(t)$,

$$(23) \quad \begin{aligned} (2k+1) & x P_k(t) P_k(x) \\ &= (k+1) P_k(t) P_{k+1}(x) + k P_k(t) P_{k-1}(x). \end{aligned}$$

This holds in particular for $k = 0$, if $P_{-1}(x)$ is arbitrarily defined in any way whatever so that $0 \cdot P_{-1} = 0$; for definiteness let $P_{-1}(x) \equiv 0$. By subtraction of (23) from the corresponding identity with t and x interchanged,

$$(2k + 1)(t - x)P_k(t)P_k(x)$$
$$= (k + 1)\left[P_{k+1}(t)P_k(x) - P_k(t)P_{k+1}(x)\right]$$
$$- k\left[P_k(t)P_{k-1}(x) - P_{k-1}(t)P_k(x)\right].$$

Summation over the values $k = 0, 1, \cdots, n$ gives

$$(t - x)\sum_{k=0}^{n}(2k + 1)P_k(t)P_k(x)$$
$$= (n + 1)\left[P_{n+1}(t)P_n(x) - P_n(t)P_{n+1}(x)\right],$$
$$K_n(x, t) = \frac{n + 1}{2} \cdot \frac{P_{n+1}(t)P_n(x) - P_n(t)P_{n+1}(x)}{t - x}.$$

This is known as *Christoffel's identity.*[*] By substitution in (22)

$$(24)\quad s_n(x) = \frac{n + 1}{2}\int_{-1}^{1}f(t)\frac{P_{n+1}(t)P_n(x) - P_n(t)P_{n+1}(x)}{t - x}dt.$$

If $f(x)$ is in particular a polynomial $\pi_n(x)$ of the nth or lower degree, $s_n(x)$ is identical with $\pi_n(x)$, according to the last paragraph of the preceding section, and (24) holds with $s_n(x)$ replaced by $\pi_n(x)$ and $f(t)$ replaced by $\pi_n(t)$. More particularly still, for $f(x) \equiv 1$, $s_n(x) \equiv 1$ for all n, and

$$(25)\quad \frac{n + 1}{2}\int_{-1}^{1}\frac{P_{n+1}(t)P_n(x) - P_n(t)P_{n+1}(x)}{t - x}dt = 1.$$

9. Solution of the differential equation. Let an expression of the form

$$(26)\qquad y = c_0 + c_1 x + c_2 x^2 + \cdots$$

[*] E. B. Christoffel, *Über die Gaussische Quadratur und eine Verallgemeinerung derselben*, Journal für die reine und angewandte Mathematik, vol. 55 (1858), pp. 61–82; p. 73.

be assumed in an application of the method of undetermined coefficients to the solution of the differential equation

(27) $(1 - x^2)y'' - 2xy' + n(n + 1)y = 0;$

the purpose is to make clear, for reference in connection with the proof of the next section, the extent to which the Legendre polynomials are characterized and identified by the fact that they satisfy (15). Substitution of (26) in the left member of (27) gives

$$(k + 2)(k + 1)c_{k+2} - \left[k(k + 1) - n(n + 1)\right]c_k$$

as the coefficient of x^k. The equation is formally satisfied if and only if

$$c_{k+2} = \frac{k(k + 1) - n(n + 1)}{(k + 1)(k + 2)} c_k.$$

Arbitrary values can be assigned to c_0 and c_1, and the rest of the c's are then successively determined, each being a determinate multiple of c_0 or c_1 according as the subscript is even or odd.

If n is a non-negative integer $c_{k+2} = 0$ when $k = n$, and it follows that $c_{n+2j} = 0$ for all positive integral values of j. For non-negative n and k the expression $k(k+1) - n(n+1)$ is different from zero when $k \neq n$, being negative if $k < n$ and positive if $k > n$. If n is even there are only a finite number of coefficients with even index which do not vanish, while all the coefficients of odd index are different from zero unless $c_1 = 0$. The differential equation, considered on its own merits, is seen to have a polynomial solution, and apart from the arbitrary constant factor c_0 has just one such solution; whatever questions of convergence may remain with regard to the infinite series obtained, the reasoning is conclusive without further analysis as far as polynomials are concerned. Similarly there is a polynomial solution for odd n, unique except for an arbitrary constant factor c_1. The Legendre polynomial is known to be a solution for each n. *If y is a polynomial in x which satisfies the differential equation* (27), *it must be a constant multiple of $P_n(x)$.*

10. Rodrigues's formula. It is to be shown that

$$(28) \qquad P_n(x) = \frac{1}{2^n n!} \frac{d^n}{dx^n} (x^2 - 1)^n.$$

This is *Rodrigues's formula*. The right-hand member, being the nth derivative of a polynomial of degree $2n$, is itself a polynomial of the nth degree. It will be shown that this polynomial is a solution of (27). A supplementary calculation will then determine the constant factor.

Let $z = (x^2 - 1)^n$. Then $z' = dz/dx = 2nx(x^2-1)^{n-1}$, so that

$$(29) \qquad (x^2 - 1)z' - 2nxz = 0.$$

By repeated differentiation of this equation

$$(x^2 - 1)z'' - (2n - 2)xz' - 2nz = 0,$$
$$(x^2 - 1)z''' - (2n - 4)xz'' - [2n + (2n - 2)]z' = 0,$$

and as a result of $k+1$ differentiations of (29)

$$(30) \quad \begin{aligned} &(x^2 - 1)z^{(k+2)} - (2n - 2k - 2)xz^{(k+1)} \\ &\quad - [2n + (2n-2) + (2n-4) + \cdots + (2n-2k)]z^{(k)} = 0. \end{aligned}$$

It would be easy, but is unnecessary for the present purpose, to write a general formula for the sum of the arithmetic progression in the last bracket. For the particular value $k = n$, however, it is to be noted that the quantity in brackets is

$$2n + (2n - 2) + \cdots + 2 + 0 = n(n + 1),$$

and the whole equation (30) becomes

$$(x^2 - 1)z^{(n+2)} + 2xz^{(n+1)} - n(n + 1)z^{(n)} = 0.$$

If $y = d^n z/dx^n$, this relation, multiplied by -1, states that

$$(1 - x^2)y'' - 2xy' + n(n + 1)y = 0.$$

Consequently, by the concluding statement of §9,

$$\frac{d^n z}{dx^n} = KP_n(x),$$

where K is independent of x.

As to the determination of K, the leading coefficient in $d^n z/dx^n$ is $2n(2n-1) \cdots (n+1)$, while the coefficient of x^n in $P_n(x)$, according to (7), is $1 \cdot 3 \cdots (2n-1)/n!$. Consequently

$$K = \frac{(2n)!}{1 \cdot 3 \cdot 5 \cdots (2n - 1)} = 2 \cdot 4 \cdot 6 \cdots (2n) = 2^n n!,$$

in agreement with (28).

11. Integral representation. Another expression for $P_n(x)$ is the integral formula

$$(31) \qquad P_n(x) = \frac{1}{\pi} \int_0^\pi [x + (x^2 - 1)^{1/2} \cos \phi]^n d\phi.$$

For the purposes of the demonstration let the right-hand member be denoted by $y_n(x)$. It is to be proved that $y_n(x)$ is identical with $P_n(x)$.

Explicit integration gives immediately $y_0(x) = 1 = P_0(x)$, $y_1(x) = x = P_1(x)$. It will be shown that $y_n(x)$ *satisfies the same relation of recurrence as* $P_n(x)$, and it will follow then that $y_2(x) = P_2(x)$, and so by repeated use of the relation that $y_n(x) = P_n(x)$ for arbitrary n.

It can be seen in advance that $y_n(x)$ is a polynomial of the nth degree (though it is not necessary for the purposes of the present section to deal with this part of the conclusion separately, as it will be covered in any event by the proof based on the recurrence formula). For if the integrand in (31) is expanded by the binomial theorem each odd power of $(x^2 - 1)^{1/2}$ is multiplied by an odd power of $\cos \phi$, the integral of which from 0 to π vanishes: by the substitution $\theta = \pi - \phi$, $\cos \theta = -\cos \phi$,

$$\int_{\pi/2}^{\pi} \cos^{2k+1} \phi \, d\phi = -(-1)^{2k+1} \int_{\pi/2}^{0} \cos^{2k+1} \theta \, d\theta$$

$$= -\int_{0}^{\pi/2} \cos^{2k+1} \theta \, d\theta,$$

so that the integral over the second half of the interval is the negative of the integral over the first half. (The fact that odd powers of $(x^2 - 1)^{1/2}$ are eliminated by the integration will be used in §12.) And each even power of $(x^2 - 1)^{1/2}$ is multiplied by an even power of $\cos \phi$, the integral of which makes a positive contribution to the coefficient of x^n, so that this coefficient can not vanish.

Let $x + (x^2 - 1)^{1/2} \cos \phi = Z$. Then

$$y_{n-1}(x) = \frac{1}{\pi} \int_{0}^{\pi} Z^{n-1} d\phi,$$

$$y_n(x) = \frac{1}{\pi} \int_{0}^{\pi} [x + (x^2 - 1)^{1/2} \cos \phi] Z^{n-1} d\phi,$$

$$y_{n+1}(x) = \frac{1}{\pi} \int_{0}^{\pi} [x + (x^2 - 1)^{1/2} \cos \phi]^2 Z^{n-1} d\phi,$$

and

$$(n + 1)y_{n+1}(x) - (2n + 1)xy_n(x) + ny_{n-1}(x)$$

(32)
$$= \frac{1}{\pi} \int_0^\pi WZ^{n-1}d\phi,$$

where

$$
\begin{aligned}
W &= (n+1)\left[x^2+2x(x^2-1)^{1/2}\cos\phi+(x^2-1)\cos^2\phi\right] \\
&\quad -(2n+1)x\left[x+(x^2-1)^{1/2}\cos\phi\right]+n \\
&= -nx^2+x(x^2-1)^{1/2}\cos\phi+(n+1)(x^2-1)\cos^2\phi+n \\
&= -n(x^2-1)(1-\cos^2\phi)+x(x^2-1)^{1/2}\cos\phi \\
&\quad +(x^2-1)\cos^2\phi \\
&= -n(x^2-1)\sin^2\phi+(x^2-1)^{1/2}\cos\phi\left[x+(x^2-1)^{1/2}\cos\phi\right].
\end{aligned}
$$

Let $-n(x^2-1)\sin^2\phi$ be denoted by U, and the rest of the last member by V, so that $W = U + V$. With Z^n and $\cos\phi\,d\phi$ as factors for an integration by parts,

$$
\begin{aligned}
\int_0^\pi VZ^{n-1}d\phi &= (x^2-1)^{1/2}\int_0^\pi Z^n\cos\phi\,d\phi \\
&= (x^2-1)^{1/2}\left\{\left[Z^n\sin\phi\right]_0^\pi \right. \\
&\quad \left. +\int_0^\pi \sin\phi\cdot nZ^{n-1}(x^2-1)^{1/2}\sin\phi\,d\phi\right\} \\
&= n(x^2-1)\int_0^\pi Z^{n-1}\sin^2\phi\,d\phi \\
&= -\int_0^\pi UZ^{n-1}d\phi.
\end{aligned}
$$

Consequently the right-hand member of (32) vanishes, and this means that the recurrence relation (5) is satisfied with $y_n(x)$ substituted for $P_n(x)$.

12. Bounds of $P_n(x)$. For values of x belonging to the interval $(-1, 1)$ let (31) be written in the form

$$(33) \quad P_n(x) = \frac{1}{\pi} \int_0^\pi [x + i(1 - x^2)^{1/2} \cos \phi]^n d\phi.$$

It was seen in §11 that odd powers of the square root, and so, in the present representation, odd powers of i, are multiplied by factors which reduce to zero when the integration is performed. That is to say, if the integrand is written in terms of its real and pure imaginary parts in the form $F_1 + iF_2$,

$$\int_0^\pi F_2 d\phi = 0, \qquad P_n(x) = \frac{1}{\pi} \int_0^\pi F_1 d\phi.$$

Since $|F_1 + iF_2| = (F_1^2 + F_2^2)^{1/2}$, and obviously $|F_1| \leqq (F_1^2 + F_2^2)^{1/2}$,

$$(34) \quad \begin{aligned} |P_n(x)| &\leqq \frac{1}{\pi} \int_0^\pi |F_1 + iF_2| \, d\phi \\ &= \frac{1}{\pi} \int_0^\pi | x + i(1 - x^2)^{1/2} \cos \phi |^n d\phi. \end{aligned}$$

(Even without the fact that $\int F_2 d\phi = 0$ it would follow from a fundamental property of integrals of complex functions that $|\int (F_1 + iF_2) d\phi| \leqq \int |F_1 + iF_2| d\phi$.) For $-1 < x < 1$,

$$\begin{aligned} | x + i(1 - x^2)^{1/2} \cos \phi | &= [x^2 + (1 - x^2) \cos^2 \phi]^{1/2} \\ &= (\cos^2 \phi + x^2 \sin^2 \phi)^{1/2} \\ &< (\cos^2 \phi + \sin^2 \phi)^{1/2} = 1 \end{aligned}$$

if $0 < \phi < \pi$; the integrand in (34) is less than 1 except at the ends of the interval of integration. Hence $|P_n(x)| < 1$ *throughout the interior of the interval* $(-1, 1)$.

A less obvious inequality for $\left| P_n(x) \right|$ over the same range will be needed for the discussion of convergence. Since $\cos^2 \phi + x^2 \sin^2 \phi = 1 - (1 - x^2) \sin^2 \phi$, it is seen by the preceding paragraph that

$$\left| P_n(x) \right| \leq \frac{1}{\pi} \int_0^\pi \left[1 - (1 - x^2) \sin^2 \phi \right]^{n/2} d\phi;$$

and as $\sin (\pi - \phi) = \sin \phi$, the integral from 0 to π is the same as twice the integral from 0 to $\pi/2$. In the latter interval, according to (39) of Chapter I, $\sin \phi \geq (2/\pi)\phi$. If $z = (2/\pi)(1 - x^2)^{1/2}$,

$$1 - (1 - x^2) \sin^2 \phi \leq 1 - (2/\pi)^2(1 - x^2)\phi^2 = 1 - z^2\phi^2.$$

By the extended mean value theorem, if y is any real number,

$$e^{-y} = 1 - y + \tfrac{1}{2}y^2 e^{-\theta y}$$

with a value of θ between 0 and 1. Since the exponential function is positive for all real values of the variable, $e^{-\theta y} > 0$, and $e^{-y} \geq 1 - y$. That is to say, $1 - y \leq e^{-y}$, and in particular $1 - z^2\phi^2 \leq e^{-z^2\phi^2}$. Consequently

$$\left| P_n(x) \right| \leq \frac{2}{\pi} \int_0^{\pi/2} e^{-nz^2\phi^2/2} d\phi.$$

With $t = n^{1/2}z\phi$ as a new variable of integration, and with $n^{1/2}z\pi/2 = A$,

$$(35) \quad \int_0^{\pi/2} e^{-nz^2\phi^2/2} d\phi = \frac{1}{n^{1/2}z} \int_0^A e^{-t^2/2} dt < \frac{1}{n^{1/2}z} \int_0^\infty e^{-t^2/2} dt.$$

With regard to the last integral all that is essential for present purposes is that it is a number independent of n and x. It follows that there is a number C independent of n and x such that

(36)
$$|P_n(x)| < \frac{C}{n^{1/2}(1 - x^2)^{1/2}}$$

for $-1 < x < 1$. *For a fixed value of x interior to $(-1, 1)$, $|P_n(x)|$ does not exceed a constant multiple of $1/n^{1/2}$.*

(With the value $(\pi/2)^{1/2}$ for the last integral in (35), the inequality (36) becomes

$$|P_n(x)| < \frac{(\pi/2)^{1/2}}{n^{1/2}(1 - x^2)^{1/2}}.$$

For an evaluation of this integral on the basis of properties of the Gamma function see the paragraph in parentheses which concludes §2 of Chapter IX.)

13. Convergence at a point of continuity interior to the interval. Let $f(x)$ be an integrable function whose square is integrable over $(-1, 1)$, and let $s_n(x)$ as in §8 be the partial sum of its development in Legendre series. By reasoning analogous to that of §7 in Chapter I,

$$\int_{-1}^{1} f(x)s_n(x)dx = \sum_{k=0}^{n} \frac{2}{2k + 1} a_k^2 = \int_{-1}^{1} [s_n(x)]^2 dx,$$

$$\int_{-1}^{1} [f(x) - s_n(x)]^2 dx = \int_{-1}^{1} [f(x)]^2 dx - \sum_{k=0}^{n} \frac{2}{2k + 1} a_k^2;$$

the left-hand member of the last equation is non-negative, the sum on the right is bounded as n increases, the corresponding infinite series is convergent, and

$$\lim_{k \to \infty} \left(\frac{2}{2k + 1}\right)^{1/2} a_k = 0,$$

$$\lim_{k \to \infty} \left(\frac{2k + 1}{2}\right)^{1/2} \int_{-1}^{1} f(x)P_k(x)dx = 0.$$

In a different notation, if $\phi(t)$ and $[\phi(t)]^2$ are integrable over $(-1, 1)$,

$$\lim_{n \to \infty} \left(\frac{2n+1}{2} \right)^{1/2} \int_{-1}^{1} \phi(t) P_n(t) dt = 0.$$

An equivalent relation may be written a little more simply with $[(2n+1)/2]^{1/2} = (n+\frac{1}{2})^{1/2}$ replaced by $n^{1/2}$, since $\lim_{n \to \infty} (n+\frac{1}{2})/n = 1$.

A fixed value interior to $(-1, 1)$ being assigned to x for a discussion of convergence, multiplication of (25) by $f(x)$ gives

$$f(x) = \frac{n+1}{2} \int_{-1}^{1} f(x) \frac{P_{n+1}(t) P_n(x) - P_n(t) P_{n+1}(x)}{t-x} \, dt.$$

If this is subtracted from (24), the result can be written with the notation

$$\phi(t) = \frac{f(t) - f(x)}{t - x}$$

in the form

$$
\begin{aligned}
(37) \quad s_n(x) - f(x) = {} & \frac{n+1}{2} P_n(x) \int_{-1}^{1} \phi(t) P_{n+1}(t) dt \\
& - \frac{n+1}{2} P_{n+1}(x) \int_{-1}^{1} \phi(t) P_n(t) dt.
\end{aligned}
$$

By §12, each of the expressions $\frac{1}{2}(n+1)P_n(x)$, $\frac{1}{2}(n+1)P_{n+1}(x)$ has an upper bound of the order of $n^{1/2}$. *If $f(x)$ is for simplicity continuous throughout the closed interval $(-1, 1)$ except for a finite number of finite jumps, the same will be true of $\phi(t)$ if $f(t)$ has a derivative at the point $t = x$, $\phi(t)$ being defined by its limit for $t = x$, or more* *generally a right-hand derivative and a left-hand derivative at the point,* whether these are the same or different, and then according to the preceding paragraph the product of either integral on the right of (37) by a quantity of the order of magnitude of $n^{1/2}$ approaches zero, so that $\lim_{n \to \infty} s_n(x) = f(x)$. *Under the hypotheses stated, the Legendre series converges at the point $t = x$ to the value $f(x)$.*

14. Convergence at a point of discontinuity interior to the interval. Let c be a number between -1 and 1, and let $f(x)$ for the moment be the particular discontinuous function which is equal to 1 for $-1 \leqq x < c$ and equal to 0 for $c < x \leqq 1$; the value assigned to the function for $x = c$ is immaterial. By §13 the Legendre series for $f(x)$ is convergent to the value $f(x)$ at any interior point of the interval other than $x = c$. The question at issue is that of convergence at the point of discontinuity.

The general coefficient in the series is

$$a_k = \frac{2k+1}{2} \int_{-1}^{c} P_k(x) dx.$$

By (14)

$$(2k+1) \int P_k(x) dx = P_{k+1}(x) - P_{k-1}(x)$$

for $k \geqq 1$, except for a constant of integration, and as $P_{k+1}(-1) = P_{k-1}(-1) = (-1)^{k-1}$,

$$a_k = \tfrac{1}{2}\big[P_{k+1}(x) - P_{k-1}(x) \big]_{-1}^{c} = \tfrac{1}{2}\big[P_{k+1}(c) - P_{k-1}(c) \big].$$

For $k = 0$ explicit integration gives

$$a_0 = \tfrac{1}{2} \int_{-1}^{c} P_0(x) dx = \tfrac{1}{2}(c+1) = \tfrac{1}{2} + \tfrac{1}{2} P_1(c).$$

The partial sum of the series for $x = c$ is

$$s_n(c) = \tfrac{1}{2} + \tfrac{1}{2}P_1(c) + \tfrac{1}{2}\sum_{k=1}^{n}\left[P_{k+1}(c) - P_{k-1}(c)\right]P_k(c)$$

$$= \tfrac{1}{2} + \tfrac{1}{2}P_{n+1}(c)P_n(c).$$

By §12, $\lim_{n\to\infty}P_n(c) = \lim_{n\to\infty}P_{n+1}(c) = 0$, and consequently

$$\lim_{n\to\infty} s_n(c) = \tfrac{1}{2};$$

the Legendre series for this particular function, like the Fourier series for a function having a similar discontinuity, converges to a value half-way between the right-hand and left-hand limits.

The result thus obtained supplies the means for its own generalization. By application of (24) to the case in hand,

$$s_n(c) = \frac{n+1}{2}\int_{-1}^{c}\frac{P_{n+1}(t)P_n(c) - P_n(t)P_{n+1}(c)}{t - c}\,dt,$$

and this approaches $\tfrac{1}{2}$ as n becomes infinite. That is to say, with x in place of c as notation for an arbitrary number interior to the interval $(-1, 1)$,

$$(38)\qquad \lim_{n\to\infty}\frac{n+1}{2}\int_{-1}^{x}\frac{P_{n+1}(t)P_n(x) - P_n(t)P_{n+1}(x)}{t - x}\,dt = \tfrac{1}{2}.$$

Furthermore, by subtraction of (38) from (25),

$$(39)\qquad \lim_{n\to\infty}\frac{n+1}{2}\int_{x}^{1}\frac{P_{n+1}(t)P_n(x) - P_n(t)P_{n+1}(x)}{t - x}\,dt = \tfrac{1}{2}.$$

Suppose now that $f(t)$ is an arbitrary function which is continuous for $-1 \leqq t \leqq 1$ except for a finite number of finite jumps, and which for a particular value of x approaches different limits $f(x+)$, $f(x-)$ as t approaches x from the right and from the left. Let it be supposed that a derivative exists on each side of the point of discontinuity, in the sense that the quotients

$$\frac{f(t) - f(x+)}{t - x}, \qquad \frac{f(t) - f(x-)}{t - x}$$

approach limits as t approaches x from the right and from the left respectively. Then a function $\phi_1(t)$ defined by the first quotient for $x < t \leq 1$ and equal to 0 in the interval from -1 to x is continuous throughout $(-1, 1)$ except for a finite number of finite jumps, and the same is true of the function $\phi_2(t)$ obtained by a complementary definition in terms of the other quotient. Let $s_n(x)$ in (24) be expressed as a sum of two parts $s_{n1}(x)$, $s_{n2}(x)$ by integrating from x to 1 for the former and from -1 to x for the latter. If (39) is multiplied by $f(x+)$ to give a representation of $\frac{1}{2}f(x+)$, and $\frac{1}{2}f(x-)$ is similarly represented by means of (38), an obvious adaptation of the reasoning of the preceding section shows that

$$\lim_{n \to \infty} s_{n1}(x) = \tfrac{1}{2}f(x+), \qquad \lim_{n \to \infty} s_{n2}(x) = \tfrac{1}{2}f(x-),$$

and consequently

$$\lim_{n \to \infty} s_n(x) = \tfrac{1}{2}\big[f(x+) + f(x-)\big].$$

It is again true that *the series converges to the mean of the limits approached from the right and from the left.*

The question of convergence at the ends of the interval, while not particularly difficult if sufficiently restrictive hypotheses are placed on $f(x)$, would require further consideration at some length, and will not be discussed here.*

SUPPLEMENTARY REFERENCES: Byerly; Stone; Churchill; Szegö; Whittaker and Watson, Chapter XV; Hobson (2); Kellogg, pp. 125–134; Courant-Hilbert; Riemann-Weber; Pólya-Szegö; Kaczmarz-Steinhaus.

* See however Ex. 11 and the accompanying discussion in the exercises on this chapter at the end of the book.

CHAPTER III

BESSEL FUNCTIONS

1. Preliminary orientation. Here, as in the preceding chapter, certain facts will be presented in logical sequence without much emphasis on motivation at the outset. An obvious bond uniting this chapter with the two preceding is the property of orthogonality of the functions concerned. An organization of ideas under which Fourier, Legendre, and Bessel series appear not merely with common characteristics, but with a common origin, will be set forth in the next two chapters, on boundary value problems. A distinguishing feature of the series to be dealt with at present, from the point of view of elementary approach, is that they involve new and specially defined functions instead of the familiar trigonometric and algebraic types.

2. Definition of $J_0(x)$. As starting point will be taken here the differential equation

$$(1) \qquad \frac{d^2y}{dx^2} + \frac{1}{x}\frac{dy}{dx} + y = 0.$$

Let a power series solution be assumed in the form

$$y = c_0 + c_1x + c_2x^2 + \cdots.$$

If this expression is substituted for y in the left-hand member of (1), the coefficient of x^k in the resulting series, *for $k \geqq 0$*, is found to be $c_k + (k+2)^2c_{k+2}$, and a necessary and sufficient condition that this coefficient vanish is that

$$(2) \qquad c_{k+2} = -\frac{c_k}{(k+2)^2}.$$

From the expansion of $(1/x)dy/dx$, however, there is a single term c_1/x, corresponding to exponent $k = -1$, which can not be cancelled by any other term, and in order that the differential equation be satisfied it is necessary that $c_1 = 0$. It follows then from (2) that c_k must be zero for every odd value of k. For even k, on the other hand, (2) is satisfied if an arbitrary value is assigned to c_0, and if

$$c_2 = -\frac{c_0}{2^2}, \qquad c_4 = -\frac{c_2}{4^2} = \frac{c_0}{2^2 4^2},$$

$$c_6 = -\frac{c_4}{6^2} = -\frac{c_0}{2^2 4^2 6^2}, \quad \cdots .$$

A formal solution of the differential equation is $c_0 J_0(x)$, where

$$(3) \qquad J_0(x) = 1 - \frac{x^2}{2^2} + \frac{x^4}{2^2 4^2} - \frac{x^6}{2^2 4^2 6^2} + \cdots ;$$

and $c_0 J_0(x)$, with arbitrary c_0, is the most general solution obtainable in this way. The result does not mean, of course, that this differential equation of the second order has essentially only one solution; it means that there is essentially only one solution *which can be expressed as a power series in x*. The point $x = 0$, at which the coefficient $1/x$ becomes infinite, is a "singular point" for the differential equation and for one of its two independent solutions.

By the ratio test, the series (3) is seen to be convergent for all values of x, the ratio of the magnitudes of the terms of degrees k and $k+2$ being $x^2/(k+2)^2$, which has the limit 0 for any fixed x as k becomes infinite. The function $J_0(x)$ is called a *Bessel function of order zero*. Bessel functions of other orders will be defined in a later section.

3. Orthogonality. Let λ be an arbitrary constant, and let $z = J_0(\lambda x)$. Then, if $t = \lambda x$, so that $z = J_0(t)$,

$$\frac{dz}{dx} = \frac{dz}{dt}\frac{dt}{dx} = \lambda J_0'(t), \qquad \frac{d^2z}{dx^2} = \frac{dt}{dx}\frac{d}{dt}\left(\frac{dz}{dx}\right) = \lambda^2 J_0''(t),$$

that is,

$$J_0'(t) = \frac{1}{\lambda}\frac{dz}{dx}, \qquad J_0''(t) = \frac{1}{\lambda^2}\frac{d^2z}{dx^2},$$

the accents indicating differentiation with respect to the accompanying variable in parentheses. Substitution of the last expressions in the identity

$$J_0''(t) + \frac{1}{t}J_0'(t) + J_0(t) = 0$$

gives

$$\frac{d^2z}{dx^2} + \frac{1}{x}\frac{dz}{dx} + \lambda^2 z = 0.$$

That is to say, z is a solution of this slightly more general differential equation.

Since $J_0(-\lambda x)$ is the same as $J_0(\lambda x)$, λ may be restricted without essential loss of generality to non-negative values.

Let λ and μ be two real numbers distinct from each other, each positive or zero, and let $J_0(\lambda x)$ now be denoted by z_1 and $J_0(\mu x)$ by z_2. These functions satisfy respectively the differential equations

$$(4) \qquad \frac{d^2 z_1}{dx^2} + \frac{1}{x}\frac{dz_1}{dx} + \lambda^2 z_1 = 0,$$

(5) $$\frac{d^2z_2}{dx^2} + \frac{1}{x}\frac{dz_2}{dx} + \mu^2 z_2 = 0.$$

If (4) multiplied by z_2 is subtracted from (5) multiplied by z_1, the result can be expressed with the notation $w = z_1 dz_2/dx - z_2 dz_1/dx$ in the form

$$dw/dx + w/x = (\lambda^2 - \mu^2)z_1 z_2.$$

Multiplied by x, the last relation becomes

$$d(wx)/dx = (\lambda^2 - \mu^2)xz_1 z_2,$$

whence by integration with respect to x from 0 to 1

(6) $$\left[x\left(z_1\frac{dz_2}{dx} - z_2\frac{dz_1}{dx} \right) \right]_0^1 = (\lambda^2 - \mu^2)\int_0^1 xz_1 z_2 dx.$$

The quantity in brackets vanishes for $x = 0$. At the other end of the interval $z_1(1) = J_0(\lambda)$, $z_2(1) = J_0(\mu)$. If λ, μ are distinct positive roots of the equation $J_0(x) = 0$,

(7) $$\int_0^1 xJ_0(\lambda x)J_0(\mu x)dx = 0.$$

This statement does not of itself assert that such numbers λ, μ exist. It will be shown in §5 that the equation $J_0(x) = 0$ has in fact infinitely many positive roots.

The content of (7) can be expressed by saying that the functions $x^{1/2}J_0(\lambda x)$, $x^{1/2}J_0(\mu x)$ are orthogonal to each other over the interval (0, 1). An alternative form of statement, having no particular advantage in the present connection, but introducing a concept which will be important later, is that $J_0(\lambda x)$ and $J_0(\mu x)$ themselves are orthogonal to each other over the interval (0, 1) *with respect to x as weight function.*

The values of dz_1/dx and dz_2/dx for $x = 1$ are $\lambda J_0'(\lambda)$ and $\mu J_0'(\mu)$. The left member of (6) vanishes if λ and μ, instead of being zeros of $J_0(x)$, are such that $J_0'(\lambda)$ $= J_0'(\mu) = 0$. More generally, if $J_0(\lambda) \neq 0$, $J_0(\mu) \neq 0$, the combination $z_1 dz_2/dx - z_2 dz_1/dx$ has for $x = 1$ the value

$$J_0(\lambda)J_0(\mu) \left[\frac{\mu J_0'(\mu)}{J_0(\mu)} - \frac{\lambda J_0'(\lambda)}{J_0(\lambda)} \right],$$

and vanishes if $xJ_0'(x)/J_0(x)$ takes on equal values for $x = \lambda$ and for $x = \mu$, i.e., if h is a constant and λ and μ satisfy the equation $xJ_0'(x)/J_0(x) = h$. *The relation of orthogonality* (7) *holds if λ and μ are distinct non-negative roots of the equation $J_0'(x) = 0$ or of the more general equation*

$$(8) \qquad\qquad xJ_0'(x) - hJ_0(x) = 0$$

(which reduces to the preceding for $h = 0$). The word *positive* is replaced by *non-negative* here because the equation $J_0'(x) = 0$ has $x = 0$ for one of its roots. The existence of infinitely many positive roots of (8) for any value of the constant h will be established in §5.

If (8) is replaced by $gxJ_0'(x) - hJ_0(x) = 0$ the earlier equation $J_0(x) = 0$ is also included as a special case, with $g = 0$.

For the evaluation of the integral in (7) when $\mu = \lambda$ let λx once more be represented by t; then

$$\int_0^1 x[J_0(\lambda x)]^2 dx = \frac{1}{\lambda^2} \int_0^\lambda t[J_0(t)]^2 dt.$$

With $y = J_0(t)$ and with accents indicating differentiation with respect to t let the differential equation

$$y'' + y'/t + y = 0$$

be multiplied by $2t^2 y'$. The result can be written in the form

$$0 = \frac{d}{dt}(t^2 y'^2) + 2t^2 yy'$$

$$= \frac{d}{dt}(t^2 y'^2) + \frac{d}{dt}(t^2 y^2) - 2ty^2,$$

whence

$$\int_0^\lambda ty^2 dt = \frac{1}{2}\left[t^2(y^2 + y'^2)\right]_0^\lambda = \frac{1}{2}\lambda^2\left\{\left[J_0(\lambda)\right]^2 + \left[J_0'(\lambda)\right]^2\right\},$$

and

$$(9) \qquad \int_0^1 x[J_0(\lambda x)]^2 dx = \frac{1}{2}\left\{[J_0(\lambda)]^2 + [J_0'(\lambda)]^2\right\}.$$

This is valid for any λ, whether satisfying one of the equations of the earlier paragraphs or not.

4. Integral representation of $J_0(x)$. A proof of the existence of the roots referred to in the preceding section is based on an integral representation:

$$(10) \qquad J_0(x) = \frac{1}{\pi}\int_0^\pi \cos(x\cos\phi)d\phi.$$

It was seen in §2 that the most general solution of the differential equation (1) in the form of a power series in x is a constant multiple of $J_0(x)$. A power series solution having 1 for its constant term must be $J_0(x)$ identically. It is to be shown that the right member of (10) is such a solution.

Let

$$y = \int_0^\pi \cos(x\cos\phi)d\phi.$$

Substitution of the series

(11) $\quad 1 - \dfrac{x^2 \cos^2 \phi}{2!} + \dfrac{x^4 \cos^4 \phi}{4!} - \dfrac{x^6 \cos^6 \phi}{6!} + \cdots$

for $\cos (x \cos \phi)$ under the sign of integration* gives

(12)
$$\frac{y}{\pi} = \frac{1}{\pi} \int_0^\pi d\phi - \frac{1}{\pi} \cdot \frac{x^2}{2} \int_0^\pi \cos^2 \phi \, d\phi + \cdots$$

$$= 1 - \frac{x^2}{4} + \cdots .$$

It would not be difficult to evaluate $\int_0^\pi \cos^{2k} \phi \, d\phi$ for arbitrary positive integral k and thus to verify term by term the identity of the series in (12) with that representing $J_0(x)$. The procedure here however, as already indicated, will be to show that y is a solution of (1).

By differentiation with respect to x under the integral sign

(13) $\qquad y' = - \displaystyle\int_0^\pi \cos \phi \sin (x \cos \phi) d\phi,$

$$y'' = - \int_0^\pi \cos^2 \phi \cos (x \cos \phi) d\phi.$$

Let integration by parts be applied in (13) with $\sin (x \cos \phi)$ and $\cos \phi \, d\phi$ as factors. It is found that

---

* From the point of view of the theory of functions the integration term by term is justified by the fact that for any fixed x the series (1) is immediately seen to be uniformly convergent for all values of ϕ. Other points of demonstration in this section and the next (and in §§9–11), such as the validity of differentiation with respect to x under the integral sign and the fact that $J_0(x)$, being sum of a power series, is continuous and therefore can not pass from a positive to a negative value without taking on the value 0, while involving considerations outside the scope of an ordinary first course in the calculus, are based with equal directness on standard theorems of analysis.

$$y' = - \left[\sin (x \cos \phi) \sin \phi\right]_0^\pi$$

$$- x \int_0^\pi \sin^2 \phi \cos (x \cos \phi)d\phi,$$

and the quantity in brackets vanishes at the ends of the interval. Thus

$$y'' + \frac{1}{x} y' + y$$
$$= \int_0^\pi (- \cos^2 \phi - \sin^2 \phi + 1) \cos (x \cos \phi)d\phi = 0.$$

It follows that y/π is identical with $J_0(x)$.

5. Zeros of $J_0(x)$ and related functions. Substitution of $\pi - \phi$ for ϕ replaces $x \cos \phi$ by $-x \cos \phi$, and leaves $\cos (x \cos \phi)$ unchanged. Consequently the integral from $\pi/2$ to π in (10) is equal to the integral from 0 to $\pi/2$, and

$$J_0(x) = \frac{2}{\pi} \int_0^{\pi/2} \cos (x \cos \phi)d\phi.$$

Let $t = x \cos \phi$, $dt = -x \sin \phi \, d\phi = -(x^2 - t^2)^{1/2}d\phi$. Then

$$(\pi/2)J_0(x) = \int_0^x \frac{\cos t}{(x^2 - t^2)^{1/2}} \, dt.$$

The denominator vanishes at the upper limit of integration, and the integrand in general becomes infinite, the improper integral being however convergent. But if x is an odd multiple of $\pi/2$ the numerator of the fraction also vanishes for $t = x$, and by the usual rule for the evaluation of the limit 0/0 it is found that the limit of the fraction for $t = x$ is 0. In the rest of the discussion of the integral attention may be restricted to such values of x.

Let $x = k\pi + \pi/2$, k being a positive integer. The graph of the integrand as a function of t crosses the t-axis at the abscissa $\pi/2$, and then consists of a succession of arches alternately below and above the axis, the last arch, from $t = (k - \frac{1}{2})\pi$ to $t = (k + \frac{1}{2})\pi$, being above the axis if k is even and below if k is odd. Let the value of the integral from 0 to $\pi/2$ be denoted by c, and let $d_1, d_2, \cdots, d_k$ denote the magnitudes of the areas between the successive arches of the curve and the axis, the integrals of the absolute value of the integrand over the intervals $(\pi/2, 3\pi/2)$, $(3\pi/2, 5\pi/2)$, $\cdots$, $(k\pi - \frac{1}{2}\pi, k\pi + \frac{1}{2}\pi)$. Then

$$(\pi/2)J_0(k\pi + \tfrac{1}{2}\pi) = c - d_1 + d_2 - d_3 + \cdots + (-1)^k d_k.$$

The values of c and of the d's individually depend on k, of course, but it is not necessary to indicate that fact explicitly in the notation.

For fixed positive k, $d_1 < d_2 < \cdots < d_k$, since the magnitudes of the ordinates of the curve at corresponding points of successive arches are the products of equal values of $|\cos t|$ by a quantity which increases from left to right. Furthermore, $c < d_1$, the integral from 0 to $\pi/2$ being smaller than the absolute value of either the integral from $\pi/2$ to π or that from π to $3\pi/2$ separately. If k is odd, therefore,

$$(\pi/2)J_0(k\pi + \tfrac{1}{2}\pi) = - (d_k - d_{k-1}) - \cdots$$
$$- (d_3 - d_2) - (d_1 - c) < 0,$$

while if k is even

$$(\pi/2)J_0(k\pi + \tfrac{1}{2}\pi) = (d_k - d_{k-1}) + \cdots$$
$$+ (d_2 - d_1) + c > 0.$$

Obviously $J_0(\tfrac{1}{2}\pi) > 0$, the integrand being positive throughout the interior of the interval in this case. So $J_0(x)$ is alternately positive and negative at the points $\pi/2$, $3\pi/2$, $5\pi/2$, $\cdots$, and *vanishes at least once in each interval between successive points of this set*. It vanishes in fact just once in each interval, but the fact that the number of roots in any one of the designated intervals can not exceed one will not be proved here. It is a consequence of general properties of functions represented by power series that $J_0(x)$ can not have more than a finite number of zeros in any finite interval; a similar remark applies to the functions $J_0'(x)$, $xJ_0'(x) - hJ_0(x)$, to be considered below.

By Rolle's theorem $J_0'(x)$ vanishes at least once in each interval between successive zeros of $J_0(x)$, and consequently *the equation $J_0'(x) = 0$ also has infinitely many positive roots*, as well as the root $x = 0$.

Before proceeding to a discussion of the more general equation (8) it is to be noted that $J_0(x)$ and $J_0'(x)$ can not vanish simultaneously. For if $J_0(a)$ and $J_0'(a)$ were both zero for $a \neq 0$ (obviously $J_0(0) = 1 \neq 0$) it would follow from the differential equation that $J_0''(a) = 0$, and then from the differentiated equation

$$J_0'''(x) + \frac{1}{x}J_0''(x) + \left(1 - \frac{1}{x^2}\right)J_0'(x) = 0$$

that $J_0'''(a) = 0$, and so by successive differentiations of the differential equation that all derivatives must vanish for $x = a$, and this would mean that $J_0(x)$ as expanded in a Taylor series of powers of $x - a$ must be identically zero.

The fact that J_0 and J_0' can not vanish simultaneously is also apparent from (9).

It is certain then that $J_0(x)$ changes sign at each point where it vanishes. If λ_1, λ_2, $\cdots$ are the positive roots of the equation $J_0(x) = 0$ in order, $J_0(x)$ changes from positive to negative values at the point λ_1, from negative to positive at λ_2, and so on: $J_0'(\lambda_1) < 0$, $J_0'(\lambda_2) > 0$, $\cdots$. But at a point where $J_0(x) = 0$, x being positive, the sign of the expression $xJ_0'(x) - hJ_0(x)$ which forms the left-hand member of (8) is that of $J_0'(x)$. So this left-hand member is alternately negative and positive at the points λ_1, λ_2, $\cdots$, and *vanishes at least once in each interval from one of these points to the next.*

6. Expansion of an arbitrary function in series. Let $\lambda_1, \lambda_2, \cdots$, be the positive roots of the equation $J_0(x) = 0$, arranged in order of increasing magnitude, or the non-negative roots of $J_0'(x) = 0$, or the positive roots of $xJ_0'(x) - hJ_0(x) = 0$ with a preassigned value of $h \neq 0$. An arbitrary function $f(x)$ on the interval $(0, 1)$ can be formally expanded in a series of the functions $J_0(\lambda_1 x)$, $J_0(\lambda_2 x)$, $\cdots$, by a procedure analogous to that followed in the case of Fourier and Legendre series.

Let an expansion be assumed in the form

$$f(x) = a_1 J_0(\lambda_1 x) + a_2 J_0(\lambda_2 x) + \cdots .$$

For the determination of the general coefficient a_k let this equation be multiplied by $xJ_0(\lambda_k x)$ and integrated from 0 to 1. By the property of orthogonality recognized in §3 the right-hand member reduces to a single term:

$$\int_0^1 xf(x)J_0(\lambda_k x)dx = a_k \int_0^1 x[J_0(\lambda_k x)]^2 dx,$$

which on evaluation of the last integral by means of (9) gives

$$a_k = \frac{2}{[J_0(\lambda_k)]^2 + [J_0'(\lambda_k)]^2} \int_0^1 xf(x)J_0(\lambda_k x)dx.$$

The denominator of course reduces to $[J_0'(\lambda_k)]^2$ if the λ's are the roots of $J_0(x) = 0$, and to $[J_0(\lambda_k)]^2$ if the roots of $J_0'(x) = 0$ are used.

No discussion of the convergence of this type of series will be undertaken here.

The formulas can be adapted by a simple change of variable to an interval $(0, a)$ of arbitrary length extending to the right from the origin.

7. Definition of $J_n(x)$. A more general differential equation including (1) as a special case is

$$(14) \qquad \frac{d^2y}{dx^2} + \frac{1}{x}\frac{dy}{dx} + \left(1 - \frac{n^2}{x^2}\right)y = 0.$$

This equation has been extensively studied for arbitrary values of n, real and complex. For present purposes however n may be thought of primarily as restricted to positive integral values (and the value 0, for which the equation reduces to (1)), though parts of the discussion will incidentally be valid without change, or with only slight changes, for arbitrary positive values or for still more general values of n.

For the sake of a slight simplification in the setting up of a power series solution let a new dependent variable be introduced by the relation $y = x^n v$. Substitution of this in (14) gives for v the differential equation

$$(15) \qquad \frac{d^2v}{dx^2} + \frac{2n+1}{x}\frac{dv}{dx} + v = 0,$$

which naturally reduces again to (1) for $n = 0$. If

$$(16) \qquad v = c_0 + c_1 x + c_2 x^2 + \cdots,$$

the coefficient of x^k when v in the left-hand member of (15) is replaced by the power series is for $k \geqq 0$

$$(k + 2)(k + 2n + 2)c_{k+2} + c_k,$$

the vanishing of which specifies that

$$c_{k+2} = - \frac{c_k}{(k + 2)(k + 2n + 2)} \; ;$$

the denominator is different from zero for all $k \geqq 0$ if $n \geqq 0$ (and more generally if n has any value except -1, $-3/2$, -2, $\cdots$). There is also a term $(2n+1)c_1/x$, for the elimination of which it is necessary that $c_1 = 0$, unless $n = -\frac{1}{2}$. For $n \geqq 0$ (in particular) c_0 is arbitrary, and the most general solution of (15) in the form (16) is c_0 times the series

$$1 - \frac{x^2}{2(2n + 2)} + \frac{x^4}{2 \cdot 4 \cdot (2n + 2)(2n + 4)}$$
$$- \frac{x^6}{2 \cdot 4 \cdot 6 \cdot (2n + 2)(2n + 4)(2n + 6)} + \cdots .$$

This series is convergent by the ratio test for all values of x.

When n is a positive integer a *Bessel function $J_n(x)$ of the nth order* is defined by choosing the constant c_0 so that $y = x^n v$ becomes

$$J_n(x) = \frac{x^n}{2^n n!} \left[1 - \frac{x^2}{2(2n + 2)} \right.$$
$$\left. + \frac{x^4}{2 \cdot 4 \cdot (2n + 2)(2n + 4)} - \cdots \right].$$
(17)

With replacement of 0! by 1 this formula reduces for $n = 0$ to the expression (3) for $J_0(x)$. On comparison of (17) for $n = 1$ with the result of differentiating (3) it is seen at once that $J_1(x) = -J_0'(x)$. For positive non-integral values of n, as well as for others to which the preceding calculation is applicable, $J_n(x)$ is defined with a corresponding value of c_0 expressed in terms of the Gamma function.*

8. Orthogonality: developments in series. With $t = \lambda x$, $z = J_n(\lambda x) = J_n(t)$, it is found by repetition of the steps of the first paragraph of §3 that

$$(18) \qquad \frac{d^2z}{dx^2} + \frac{1}{x} \frac{dz}{dx} + \left(\lambda^2 - \frac{n^2}{x^2}\right)z = 0.$$

For the purposes of the text it is to be understood throughout that n is a positive integer (or 0, the conclusions for $n = 0$ being merely a reiteration of those already obtained). The reasoning of this paragraph and the next two is equally valid to be sure for arbitrary positive n; it is independent of the specification of c_0 in the last sentence of the preceding section. In considering the possibility of extension to negative values of n it is to be borne in mind that when n is negative $J_n(x)$ in general becomes infinite for $x = 0$.

When the differential equations satisfied by $z_1 = J_n(\lambda x)$, $z_2 = J_n(\mu x)$ are combined in accordance with the procedure of the third paragraph of §3, the terms involving n^2 cancel each other, and the formulas leading up to the relation of orthogonality are from that point on exactly the same as before, except for the replacement of J_0 by J_n. *If λ and μ are distinct positive roots of the equation* $J_n(x) = 0$, *or of the equation* $J_n'(x) = 0$, *or of the equation* $xJ_n'(x) - hJ_n(x) = 0$,

$$\int_0^1 xJ_n(\lambda x)J_n(\mu x)dx = 0.$$

* See Exs. 4–13 and the accompanying discussion among the exercises on this chapter at the end of the book.

Explicit admission of the value 0 for λ or μ in this statement would be trivial, since $J_n(0 \cdot x) \equiv 0$ for $n > 0$.

For the evaluation of the last integral when $\mu = \lambda$ the procedure of the concluding paragraph of §3 is again effective. The term involving n^2 is in evidence this time throughout the calculation, and (9) is generalized to the form

(19)
$$\int_0^1 x[J_n(\lambda x)]^2 dx$$
$$= \frac{1}{2}\left\{[J_n'(\lambda)]^2 + \left(1 - \frac{n^2}{\lambda^2}\right)[J_n(\lambda)]^2\right\}.$$

The existence of the roots on which the significance of the statement of orthogonality depends will be proved in §11.

If $\lambda_1, \lambda_2, \cdots$, are the positive roots of the equation $J_n(x) = 0$ or of the equation $J_n'(x) = 0$ (n being a positive integer) there corresponds to any integrable function $f(x)$ on the interval $(0, 1)$ an expansion (whether convergent or not) of the form

$$a_1 J_n(\lambda_1 x) + a_2 J_n(\lambda_2 x) + \cdots,$$

with

$$a_k = \frac{\int_0^1 x f(x) J_n(\lambda_k x) dx}{\int_0^1 x[J_n(\lambda_k x)]^2 dx},$$

in which the denominator is to be evaluated by means of (19). The same is true when the λ's are the roots of $x J_n'(x) - h J_n(x) = 0$, except that an additional term is needed in a particular case to make the series complete, namely when $h = n$; since in this case the function $z = x^n$ is a solution of (18) with $\lambda = 0$, and at the same time satisfies the condition that $dz/dx = hz$ for $x = 1$, adaptation of the calculation of the third paragraph of §3 with

$z_1 = x^n$, $z_2 = J_n(\lambda_k x)$ shows that x^n is orthogonal to each of the functions $J_n(\lambda_k x)$ with respect to x as weight function.*

9. Integral representation of $J_n(x)$. A representation of $J_n(x)$ corresponding to (10) for positive integral n is

$$(20) \qquad \frac{1}{\pi} \cdot \frac{x^n}{1 \cdot 3 \cdot 5 \cdots (2n-1)} \int_0^\pi \sin^{2n} \phi \cos (x \cos \phi) d\phi.$$

Let the integral in this formula be denoted by v. On substitution of (11) for $\cos (x \cos \phi)$ it becomes apparent that v is represented by a power series in x with constant term

$$(21) \qquad A_n = \int_0^\pi \sin^{2n} \phi \, d\phi.$$

Differentiation with respect to x under the integral sign gives

$$(22) \qquad \frac{dv}{dx} = - \int_0^\pi \sin^{2n} \phi \cos \phi \sin (x \cos \phi) d\phi,$$

$$\frac{d^2v}{dx^2} = - \int_0^\pi \sin^{2n} \phi \cos^2 \phi \cos (x \cos \phi) d\phi.$$

Let $\sin (x \cos \phi)$ and $\sin^{2n} \phi \cos \phi \, d\phi$ be used as factors for an integration by parts in (22). Since the function $\sin^{2n+1}\phi/(2n+1)$ obtained by integration of the second factor vanishes at the ends of the interval,

$$(23) \qquad \frac{dv}{dx} = - \frac{x}{2n+1} \int_0^\pi \sin^{2n+2} \phi \cos (x \cos \phi) d\phi.$$

* See e.g. C. N. Moore, *The summability of the developments in Bessel functions, with applications*, Transactions of the American Mathematical Society, vol. 10 (1909), pp. 391–435; pp. 418–419.

Here $\sin^{2n+2}\phi = \sin^{2n}\phi \sin^2\phi$, and it is seen that v satisfies (15). Consequently, by the discussion of this differential equation in §7, v is a constant multiple of $J_n(x)/x^n$. It remains to verify the correctness of the constant factor in (20).

Successive changes of the variable of integration in (21), using first the supplementary angle, then the complementary angle, then the supplementary angle again, show that

$$A_n = 2 \int_0^{\pi/2} \sin^{2n} \phi \, d\phi = 2 \int_0^{\pi/2} \cos^{2n} \phi \, d\phi = \int_0^{\pi} \cos^{2n} \phi \, d\phi.$$

From the last form of representation it is recognized on substitution of (11) in (10) that $(-1)^n A_n/[\pi(2n)!]$ is the coefficient of x^{2n} in the power series for $J_0(x)$, or by comparison with (3)

$$\frac{A_n}{(2n)!\pi} = \frac{1}{2^2 4^2 \cdots (2n)^2}, \qquad A_n = \frac{1 \cdot 3 \cdot 5 \cdots (2n-1)}{2 \cdot 4 \cdot 6 \cdots (2n)} \pi.$$

According to (17), $J_n(x)/x^n$ is a power series with $1/(2^n n!)$ as constant term, and consequently

$$v = 1 \cdot 3 \cdot 5 \cdots (2n-1)\pi J_n(x)/x^n,$$

in agreement with (20).

10. Recurrence formulas. Let the integral v of the preceding section now be denoted by v_n. Then according to (23)

$$\frac{dv_n}{dx} = -\frac{xv_{n+1}}{2n+1}.$$

Let $1 \cdot 3 \cdot 5 \cdots (2n-1)\pi = C_n$. As $v_n = C_n J_n(x)/x^n$ and $(2n+1)C_n = C_{n+1}$ it follows that

$$\frac{d}{dx}\left(\frac{J_n(x)}{x^n}\right) = \frac{1}{C_n}\frac{dv_n}{dx} = -\frac{xv_{n+1}}{C_{n+1}} = -\frac{J_{n+1}(x)}{x^n},$$

whence

$$(24) \qquad J_{n+1}(x) = \frac{n}{x}J_n(x) - J_n'(x).$$

For positive integral n an alternative integration by parts may be performed in (22) with $\sin^{2n-1}\phi\cos\phi$ and

$$\sin\phi\sin(x\cos\phi)d\phi = (1/x)d\cos(x\cos\phi)$$

as factors. Since

$$(d/d\phi)\sin^{2n-1}\phi\cos\phi = (2n-1)\sin^{2n-2}\phi\cos^2\phi - \sin^{2n}\phi$$

$$= (2n-1)\sin^{2n-2}\phi - 2n\sin^{2n}\phi$$

it is found that

$$\frac{dv_n}{dx} = \frac{1}{x}\left[(2n-1)v_{n-1} - 2nv_n\right].$$

Consequently

$$\frac{d}{dx}\left[x^nJ_n(x)\right] = \frac{d}{dx}(x^{2n}v_n/C_n) = (2n-1)x^{2n-1}v_{n-1}/C_n$$

$$= x^{2n-1}v_{n-1}/C_{n-1} = x^nJ_{n-1}(x),$$

and

$$(25) \qquad J_{n-1}(x) = \frac{n}{x}J_n(x) + J_n'(x).$$

Subtraction and addition of (24) and (25) give

$$J_n'(x) = \tfrac{1}{2}\left[J_{n-1}(x) - J_{n+1}(x)\right],$$

$$J_{n+1}(x) + J_{n-1}(x) = (2n/x)J_n(x).$$

The last is a relation of recurrence by means of which $J_n(x)$ for any integral $n \geqq 2$ can be expressed in terms of J_0 and J_1, and so in terms of J_0 and J_0'.

11. Zeros. By reasoning similar to that applied to $J_0(x)$ in the latter part of §5, and based on successive differentiations of the differential equation, or, alternatively, on the relation (19), it can be seen that $J_n(x)$ and $J_n'(x)$ do not vanish simultaneously for any value of $x \neq 0$. (While $J_0(0) \neq 0$ and $J_1'(0) \neq 0$, it is apparent that $J_n(0) = J_n'(0) = 0$ for $n \geqq 2$.)

At any two successive positive roots of the equation $J_n(x) = 0$ the values of $J_n'(x)$, and so the values taken on by the right-hand member of (24), have opposite signs, and $J_{n+1}(x)$ therefore must vanish at an intermediate point. If $J_n(x)$ vanishes for infinitely many positive values of x the same must be true of $J_{n+1}(x)$. Since the existence of infinitely many zeros has been proved for J_0 (and for $J_1 = -J_0'$), it follows by induction for arbitrary positive integral n.

Transition from the equation $J_n(x) = 0$ to the equations $J_n'(x) = 0$, $xJ_n'(x) - hJ_n(x) = 0$ can then be made by the same arguments that were used in §5 for $n = 0$.

12. Asymptotic formula. The general character of the Bessel functions is made much more clearly apparent, and important information is supplied in preparation for the study of convergence of Bessel series, by "asymptotic formulas" showing the behavior of the functions for large values of the variable. The general theory of such asymptotic formulas is by no means simple. On the basis of the fact that the function $u = x^{1/2}J_n(x)$ satisfies the differential equation

$$u'' + [1 - (n^2 - \tfrac{1}{4})/x^2]u = 0$$

it is however easy to show that for fixed n there are constants α, δ such that

$$J_n(x) = \frac{\alpha}{x^{1/2}} \sin (x + \delta) + \frac{r_n(x)}{x^{3/2}},$$

where $r_n(x)$ is a function which remains bounded as x becomes infinite.*

13. Orthogonal functions arising from linear boundary value problems. The functions $\cos \lambda x$ and $\sin \lambda x$ are solutions of the differential equation

$$(26) \qquad\qquad y'' + \lambda^2 y = 0.$$

They satisfy the further conditions

$$(27) \qquad y(-\pi) = y(\pi), \qquad y'(-\pi) = y'(\pi)$$

if and only if λ is an integer. That is to say, the functions 1, $\cos x$, $\sin x$, $\cos 2x$, $\sin 2x$, $\cdots$, from which a Fourier series is formed, can be regarded as solutions of the *boundary value problem* defined by the differential equation (26) and the auxiliary conditions (27) for the interval $(-\pi, \pi)$.

The differential equation of the Legendre polynomials has the points $x = \pm 1$ as *singular points*, in the sense that when it is written in the form

$$y'' - \frac{2x}{1 - x^2} y' + \frac{n(n + 1)}{1 - x^2} y = 0,$$

with the coefficient of y'' reduced to unity, the other coefficients become infinite at the points in question. The solutions $P_n(x)$, corresponding to integral values of n, are distinguished by the fact that they remain finite

* See Courant-Hilbert, pp. 286–288

at the singular points. They can be regarded as solutions of a boundary value problem, relating in this case to the interval $(-1, 1)$, with auxiliary conditions of a somewhat different type from those considered in the last paragraph.

The equation

$$y'' + (1/x)y' + y = 0$$

has a singular point for $x = 0$. The solutions used in forming the series of §6 remain finite at the singular point, and satisfy a further auxiliary condition at the other end of the interval $(0, 1)$.

Thus the series of Fourier, Legendre, and Bessel, already recognized as possessing a fundamental resemblance, are brought together in another way under a common classification.

A more general problem of essentially similar character, though it does not reduce precisely to any of those just mentioned as a special case, is associated with the differential equation

$$(28) \qquad y'' + [\lambda^2 - \phi(x)]y = 0$$

and a pair of boundary conditions of the form

$$(29) \qquad y'(0) - hy(0) = 0, \qquad y'(\pi) + Hy(\pi) = 0,$$

in which $\phi(x)$ is a given function continuous over the interval to be considered, taken for definiteness as the interval from 0 to π, and h and H are constants. The specifications (28), (29) define a *Sturm-Liouville* boundary value problem. Non-trivial solutions exist only for a certain infinite set of isolated values of the parameter λ, called *characteristic numbers*; the corresponding solutions are called *characteristic* solutions or *characteristic functions*. The functions 1, cos x, cos $2x$, $\cdots$, which make up a Fourier cosine series, as distinguished from a complete Fourier series, are given by a differential system of exactly this form, with $\phi(x) \equiv 0$, $h = H = 0$. The functions $\sin kx$, $k = 1, 2, \cdots$, without the cosines, are obtained with $\phi(x) \equiv 0$ if the conditions (29) are replaced by $y(0) = y(\pi) = 0$.

If y_1 and y_2 are characteristic solutions of (28) and (29) belonging to parameter values λ_1, λ_2, it is found by combining the differential equations satisfied by y_1 and y_2 respectively that

$$\frac{d}{dx}(y_1 y_2' - y_2 y_1') = (\lambda_1^2 - \lambda_2^2) y_1 y_2,$$

$$[y_1 y_2' - y_2 y_1']_0^\pi = (\lambda_1^2 - \lambda_2^2) \int_0^\pi y_1 y_2 \, dx.$$

Since the left member of the last equation vanishes, by (29), the functions y_1, y_2 are orthogonal to each other if $\lambda_1^2 \neq \lambda_2^2$. The infinite sequence of characteristic functions can be used for the formal development of an arbitrary function in series. (See reference to Ince.)

Still more general boundary value and expansion problems arise in association with linear homogeneous differential equations of arbitrary order. (See reference to Birkhoff.)

The notion of boundary value problems is met with again in the next two chapters, in connection with certain partial differential equations.

SUPPLEMENTARY REFERENCES: Byerly; Churchill; Whittaker and Watson, Chapter XVII; Gray, Mathews, and MacRobert; Watson; Courant-Hilbert; Riemann-Weber; for §13: Ince; Birkhoff.

CHAPTER IV

BOUNDARY VALUE PROBLEMS

1. Fourier series: Laplace's equation in an infinite strip. The series of Fourier, Legendre, and Bessel, together with others, have a common field of application in connection with the solution of what are commonly called the "partial differential equations of mathematical physics." One of the most important of these is *Laplace's equation*, having for three independent variables the form

$$(1) \qquad \frac{\partial^2 u}{\partial x^2} + \frac{\partial^2 u}{\partial y^2} + \frac{\partial^2 u}{\partial z^2} = 0.$$

No less important is the corresponding equation in two independent variables,

$$(2) \qquad \frac{\partial^2 u}{\partial x^2} + \frac{\partial^2 u}{\partial y^2} = 0,$$

to which (1) reduces either if a plane problem is under consideration instead of one in space, or if the functions relating to a space problem are so specialized as to be independent of the z-coordinate. Equation (2) also has a significance of its own in the theory of functions of a complex variable.

The question to be discussed is not merely that of finding functions which will satisfy the differential equations. It is easy in fact to write down formally a general solution of (2), namely

$$\phi_1(x + iy) + \phi_2(x - iy),$$

91

where ϕ_1 and ϕ_2 are arbitrary functions of a single variable having the requisite derivatives, and i is the imaginary unit. The problem is to select from the multiplicity of solutions a particular one to serve a specific purpose.

Let it be required as an illustration to find a solution of (2) defined on a strip of the (x, y)-plane bounded by the lines $x = 0$, $x = \pi$, and $y = 0$, and satisfying, in addition to the differential equation, the boundary conditions

$$(3) \qquad\qquad u(x, 0) = f(x),$$

$$(4) \qquad\qquad u(0, y) = 0,$$

$$(5) \qquad\qquad u(\pi, y) = 0,$$

$$(6) \qquad\qquad \lim_{y \to +\infty} u(x, y) = 0,$$

where $f(x)$ is a given function assigned in advance. With reference to an idealized physical problem, u may then be regarded as the temperature at the point (x, y) in an infinite strip of conducting material having the specified boundaries, if temperatures represented from point to point by the function $f(x)$ are maintained on the end of the strip, the long edges are kept at temperature zero, and the temperature approaches zero in the distant part of the strip, the flow of heat being supposed to have reached a "steady state" so that the temperature at any fixed point is independent of the time. A problem closer to physical reality, but a little less simple as to its mathematical solution, in which the infinite strip is replaced by a finite rectangle, will be discussed in the next section.

The designation of the width of the strip as π, for the sake of simplifying the subsequent notation, is not to be thought of as an artificial limitation on the generality of the physical conditions, but merely as amounting to a legitimate and convenient choice of the unit of linear measurement.

A device frequently used consists in seeking a solution of the differential equation first in the special form $X(x) Y(y)$, where each factor is a function of just one of the variables. Substitution of the product XY for u in (2) leads to the equation $Y''/Y = -X''/X$. Since the two members of this equation are required to represent the same quantity, which by the first representation is independent of x and by the second is independent of y, *this quantity must be a constant.* As far as the differential equation is concerned the constant may be positive or negative; for the problem in hand it turns out to be convenient to take it as positive or zero, and to represent it by λ^2.

Then X and Y separately satisfy the ordinary differential equations

$$X''(x) = -\lambda^2 X(x), \qquad Y''(y) = \lambda^2 Y(y),$$

with the same value of λ in both. The first has the solutions $\cos \lambda x$, $\sin \lambda x$; the second has the solutions $e^{\lambda y}$, $e^{-\lambda y}$. Combination of these gives for (2) the particular solutions

$$(7) \quad e^{\lambda y} \cos \lambda x, \quad e^{\lambda y} \sin \lambda x, \quad e^{-\lambda y} \cos \lambda x, \quad e^{-\lambda y} \sin \lambda x,$$

each of which is found by substitution actually to satisfy the differential equation for arbitrary λ. With sufficient generality for present purposes, and in fact, as is obvious, without any material loss of generality at all, it may be supposed that $\lambda \geq 0$.

Of the four functions in (7), the first two become infinite as y becomes positively infinite, while each of the others satisfies the auxiliary condition (6). The function $e^{-\lambda y} \sin \lambda x$, still with arbitrary $\lambda \geqq 0$, satisfies (4) as well as (2) and (6). It satisfies (5) if λ is any *integer*. The requirements (2), (4), (5) and (6) then are fulfilled by each of the functions $e^{-ny} \sin nx$, $n = 1, 2, 3, \cdots$.

Since the conditions thus far satisfied are homogeneous, containing only terms which are of the first degree as to their dependence on the unknown function and its derivatives, any constant multiple of a solution is a solution, and the sum of any two solutions is a solution. It follows then that (2), (4), (5) and (6) are satisfied by the sum of any finite number of terms of the form $b_n e^{-ny} \sin nx$. It is naturally to be anticipated, and can be proved by routine methods of analysis which will not be presented here, that under suitable conditions of convergence the same is true of an infinite series of such terms. The formal solution of the boundary value problem is completed by setting

$$(8) \quad \begin{aligned} u(x, y) = {} & b_1 e^{-y} \sin x + b_2 e^{-2y} \sin 2x \\ & + b_3 e^{-3y} \sin 3x + \cdots , \end{aligned}$$

and determining the coefficients so that

$$f(x) = u(x, 0) = b_1 \sin x + b_2 \sin 2x + b_3 \sin 3x + \cdots ,$$

i.e., by using in (8) the coefficients of the Fourier sine series for $f(x)$.

If the width of the strip is a instead of π, the condition (5) being replaced by $u(a, y) = 0$, the notation is complicated merely to the extent that $e^{-n\pi y/a} \sin (n\pi x/a)$ is to be substituted for $e^{-ny} \sin nx$, and $f(x)$ is to be expanded in a series $\sum b_n \sin (n\pi x/a)$ on the interval $(0, a)$, the generalization referred to in the next to the last paragraph of §3 in Chapter I being applied here to the sine series without cosine terms.

2. Fourier series: Laplace's equation in a rectangle.
A problem susceptible of more complete physical realization is concerned with a rectangle instead of an infinite strip. Let b denote the length of the rectangle, the width being taken again for the time being as π. A solution of (2) is to be found satisfying the conditions (3), (4), (5) and

$$(9) \qquad u(x, b) = 0.$$

(The discussion already given is physically significant in the sense that it gives a close approximation to the solution of the present problem if b is even moderately large, because of the rapid diminution of the exponential factors as y increases.)

The functions $e^{ny} \sin nx$ and $e^{-ny} \sin nx$ satisfy (2), (4), and (5) for arbitrary positive integral n. The combination

$$\tfrac{1}{2}e^{nb}e^{-ny} \sin nx - \tfrac{1}{2}e^{-nb}e^{ny} \sin nx = \sinh n(b - y) \sin nx$$

also satisfies (9). The function

$$u(x, y) = \sum_{n=1}^{\infty} c_n \sinh n(b - y) \sin nx$$

satisfies formally all the conditions of the problem if the coefficients are such that

$$f(x) = u(x, 0) = c_1 \sinh b \sin x + c_2 \sinh 2b \sin 2x + \cdots;$$

c_n is the corresponding coefficient in the sine series for $f(x)$, divided by $\sinh nb$.

The solution can of course be adapted to a rectangle of width a instead of π. It is clear also that by a slight modification in the details of the work, or a simple change of variable in the result, the formulas can be made to fit arbitrarily given boundary values on the upper instead of the lower end of the rectangle. By interchange of x and y, with a corresponding interchange of a and b, they become applicable to boundary values given on a vertical side. Let $u_1(x, y), \cdots, u_4(x, y)$ be solutions of (2) satisfying respectively the following sets of boundary conditions:

$$u_1(x, 0) = f_1(x), \quad u_1(x, b) = 0,$$
$$u_1(0, y) = 0, \quad u_1(a, y) = 0,$$
$$u_2(x, 0) = 0, \quad u_2(x, b) = f_2(x),$$
$$u_2(0, y) = 0, \quad u_2(a, y) = 0,$$
$$u_3(x, 0) = 0, \quad u_3(x, b) = 0,$$
$$u_3(0, y) = f_3(y), \quad u_3(a, y) = 0,$$
$$u_4(x, 0) = 0, \quad u_4(x, b) = 0,$$
$$u_4(0, y) = 0, \quad u_4(a, y) = f_4(y),$$

and let $u(x, y) = u_1 + u_2 + u_3 + u_4$. Then

$$u(x, 0) = f_1(x), \quad u(x, b) = f_2(x),$$
$$u(0, y) = f_3(y), \quad u(a, y) = f_4(y),$$

that is to say, $u(x, y)$ is a solution of Laplace's equation taking on boundary values arbitrarily assigned on the entire perimeter of the rectangle.

In a general theory of the properties of functions satisfying Laplace's equation, commonly called *potential theory*, it is shown *a priori*, not merely for a rectangle but for much more general regions, that the equation has one and just one solution taking on given boundary values.

3. Fourier series: vibrating string. The vibrations of a stretched uniform elastic string fastened at the ends are described, under idealized conditions representing a close approximation to physical reality, by a partial differential equation of the second order, in connection with which Fourier series can be employed again to construct the required solution.

Let an x-axis be taken through the fixed ends of the string, with the origin at one end. Let it be supposed that the motion is all in one plane, and let this be taken as the (x, y)-plane. If the point whose equilibrium position would be $(x, 0)$ is displaced at the instant t to the position (x, y), the motion is completely described in mathematical terms by determining y as a function of x and t. This function $y(x, t)$ satisfies the differential equation

$$(10) \qquad \frac{\partial^2 y}{\partial t^2} = a^2 \frac{\partial^2 y}{\partial x^2},$$

where a (taken as positive) is a constant depending on the units of measurement and the physical properties of the string. For convenience, let the unit of length be chosen so that the length of the string is π.

The equation has the general solution

$$y(x, t) = \phi_1(x + at) + \phi_2(x - at),$$

where ϕ_1 and ϕ_2 are arbitrary functions of a single variable. A question calling for more detailed examination is once more that of specifying a solution which meets the requirements of a particular problem.

Let the motion be started by distorting the string into the shape $y = f(x)$, and releasing it from rest when $t = 0$. By virtue of these initial conditions and the fact that the ends are fixed, $y(x, t)$ is to satisfy, together with (10), the relations

$$(11) \qquad\qquad y(0, t) = 0,$$

(12) $y(\pi, t) = 0,$

(13) $y(x, 0) = f(x),$

(14) $y_t(x, 0) = 0,$

y_t denoting $\partial y/\partial t$.

Particular solutions of (10) may be found by restricting y at first to the form $X(x) \, T(t)$. Then

(15) $T''/(a^2T) = X''/X.$

This quantity, being independent of x in consequence of the first form of representation and independent of t by the second, is a constant, which may be denoted by $-\lambda^2$; positive values of it would not be useful in the solution of the present problem. The ordinary differential equations

(16) $T''(t) = -\lambda^2 a^2 T(t),$ $X''(x) = -\lambda^2 X(x)$

for T and X respectively have the solutions

$$\cos \lambda at, \quad \sin \lambda at; \quad \cos \lambda x, \quad \sin \lambda x,$$

and (10) is satisfied by

$$\cos \lambda x \cos \lambda at, \ \cos \lambda x \sin \lambda at, \ \sin \lambda x \cos \lambda at, \ \sin \lambda x \sin \lambda at.$$

It may be supposed without loss of generality that $\lambda \geqq 0$.

Of these expressions the first and third, containing the factor $\cos \lambda at$, satisfy (14). The product $\sin \lambda x \cos \lambda at$ satisfies (11) for all λ, and satisfies (12) if λ is an integer. A function satisfying the differential equation and all the auxiliary conditions is constructed by making

(17) $y(x, t) = \sum_{n=1}^{\infty} b_n \sin nx \cos nat,$

$$f(x) = y(x, 0) = \sum_{n=1}^{\infty} b_n \sin nx.$$

Again the concluding step is the development of a given function in a sine series.

If in particular $f(x) = h \sin x$, where h is a constant, i.e., if the initial shape of the string is a single arch of a sine curve, the sine series for $f(x)$ reduces to the single term which represents $f(x)$ identically, and the corresponding expression for y is $h \sin x \cos at$. At any instant t the shape of the string is described by the same function $\sin x$, multiplied by a factor depending on t. The point of the string corresponding to any fixed x, on the other hand, performs a simple harmonic motion with amplitude $h \sin x$ and period $2\pi/a$. This motion of the string in the form of a single sine arch is called its *fundamental* vibration, and the corresponding musical tone is called the fundamental tone of the string.

If $f(x) = h \sin 2x$, then $y = h \sin 2x \cos 2at$. For any t the string (if not in instantaneous coincidence with the x-axis) has the form of two arches of a sine curve, the middle point $x = \pi/2$, called under these circumstances a *node*, remaining motionless in the equilibrium position. For fixed x the motion is simple harmonic with period $2\pi/(2a)$. This vibration, or the musical tone which it produces, is called the first *harmonic* or *overtone*. Higher harmonics are similarly defined, the frequencies (reciprocals of the periods) being successive multiples of the fundamental frequency. The general oscillation (17) can be regarded as resulting from superposition of the fundamental and the various harmonics with suitable amplitude factors.

The above problem is spoken of as that of a *plucked* string. If the string is *struck* instead of plucked, it is initially in the equilibrium position, and its particles are given initial transverse velocities of assigned magnitude, varying in general from point to point. The conditions (13), (14) are to be replaced by

$$y(x, 0) = 0,$$

(18) $$y_t(x, 0) = \phi(x).$$

A particular function meeting all the conditions except (18) is $\sin nx \sin nat$. The final steps of the solution are described by the formulas

$$y(x, t) = \sum_{n=1}^{\infty} c_n \sin nx \sin nat,$$

$$y_t(x, t) = \sum_{n=1}^{\infty} nac_n \sin nx \cos nat,$$

$$\phi(x) = y_t(x, 0) = \sum_{n=1}^{\infty} nac_n \sin nx.$$

The resultant oscillation is made up of a fundamental and harmonics of the same frequencies as before.

4. Fourier series: damped vibrating string. If the motion of the string is subject to a resistance proportional to the velocity, the differential equation (10) is replaced by

$$\frac{\partial^2 y}{\partial t^2} + k \frac{\partial y}{\partial t} = a^2 \frac{\partial^2 y}{\partial x^2},$$

and (15) in conjunction with (16) by

$$(T'' + kT')/(a^2 T) = X''/X = -\lambda^2.$$

The equation for X is the same as before. Particular solutions for T are

$$e^{-kt/2} \cos \mu t, \qquad e^{-kt/2} \sin \mu t, \qquad \mu = \tfrac{1}{2}(4\lambda^2 a^2 - k^2)^{1/2},$$

if k is small enough so that $k^2 - 4\lambda^2 a^2 < 0$.

Auxiliary conditions are the same as when there is no damping. In the case of the plucked string, solutions for all the conditions except (13) are given by

$$e^{-kt/2}\left(\cos \mu_n t + \frac{k}{2\mu_n} \sin \mu_n t\right) \sin nx, \qquad n = 1, 2, \cdots,$$

$$\mu_n = \tfrac{1}{2}(4n^2a^2 - k^2)^{1/2} = na\left(1 - \frac{k^2}{4n^2a^2}\right)^{1/2}.$$

For agreement with (13) also,

$$y(x, t) = \sum_{n=1}^{\infty} b_n e^{-kt/2}\left(\cos \mu_n t + \frac{k}{2\mu_n} \sin \mu_n t\right) \sin nx,$$

$$f(x) = \sum_{n=1}^{\infty} b_n \sin nx.$$

The fact that the frequencies $\mu_n/(2\pi)$ corresponding to the factors $\cos \mu_n t$, $\sin \mu_n t$ are not (in general) simply related to each other is associated with a quality of dissonance in the sound given out by the vibrating string, if the damping is appreciable.

5. Polar coordinates in the plane. For some problems there is occasion to transform the equation (2) to polar coordinates. Let such coordinates (r, ϕ) be related to the rectangular coordinates (x, y) by the equations

$$x = r \cos \phi, \qquad y = r \sin \phi.$$

If u is a function of the coordinates,

$$\frac{\partial u}{\partial r} = \cos \phi \frac{\partial u}{\partial x} + \sin \phi \frac{\partial u}{\partial y},$$

$$\frac{\partial u}{\partial \phi} = -r \sin \phi \frac{\partial u}{\partial x} + r \cos \phi \frac{\partial u}{\partial y}.$$

Solution of these linear equations for $\partial u/\partial x$ and $\partial u/\partial y$ gives

$$(19) \qquad \frac{\partial u}{\partial x} = \cos \phi \, \frac{\partial u}{\partial r} - \frac{1}{r} \sin \phi \, \frac{\partial u}{\partial \phi},$$

$$\frac{\partial u}{\partial y} = \sin \phi \, \frac{\partial u}{\partial r} + \frac{1}{r} \cos \phi \, \frac{\partial u}{\partial \phi}.$$

Let (19) be written with $\partial u/\partial x$ in place of u:

$$\frac{\partial}{\partial x}\left(\frac{\partial u}{\partial x}\right) = \cos \phi \, \frac{\partial}{\partial r}\left(\frac{\partial u}{\partial x}\right) - \frac{1}{r} \sin \phi \, \frac{\partial}{\partial \phi}\left(\frac{\partial u}{\partial x}\right).$$

By carrying out in detail the indicated operations of differentiation with respect to r and ϕ on $\partial u/\partial x$ as given by (19) it is found that

$$\frac{\partial^2 u}{\partial x^2} = \cos \phi \left[\cos \phi \, \frac{\partial^2 u}{\partial r^2} + \frac{1}{r^2} \sin \phi \, \frac{\partial u}{\partial \phi} - \frac{1}{r} \sin \phi \, \frac{\partial^2 u}{\partial r \partial \phi} \right]$$

$$- \frac{1}{r} \sin \phi \left[- \sin \phi \, \frac{\partial u}{\partial r} + \cos \phi \, \frac{\partial^2 u}{\partial r \partial \phi} \right.$$

$$\left. - \frac{1}{r} \cos \phi \, \frac{\partial u}{\partial \phi} - \frac{1}{r} \sin \phi \, \frac{\partial^2 u}{\partial \phi^2} \right].$$

The fact that some of these terms can be combined is not important for the moment. Similarly,

$$\frac{\partial^2 u}{\partial y^2} = \sin \phi \left[\sin \phi \, \frac{\partial^2 u}{\partial r^2} - \frac{1}{r^2} \cos \phi \, \frac{\partial u}{\partial \phi} + \frac{1}{r} \cos \phi \, \frac{\partial^2 u}{\partial r \partial \phi} \right]$$

$$+ \frac{1}{r} \cos \phi \left[\cos \phi \, \frac{\partial u}{\partial r} + \sin \phi \, \frac{\partial^2 u}{\partial r \partial \phi} \right.$$

$$\left. - \frac{1}{r} \sin \phi \, \frac{\partial u}{\partial \phi} + \frac{1}{r} \cos \phi \, \frac{\partial^2 u}{\partial \phi^2} \right].$$

Addition of these expressions gives

(20) $$\frac{\partial^2 u}{\partial x^2} + \frac{\partial^2 u}{\partial y^2} = \frac{\partial^2 u}{\partial r^2} + \frac{1}{r}\ \frac{\partial u}{\partial r} + \frac{1}{r^2}\ \frac{\partial^2 u}{\partial \phi^2}\ .$$

The same result can be found with less calculation by a process involving application of Green's theorem, which however is outside the scope of the present treatment.

6. Fourier series: Laplace's equation in a circle; Poisson's integral. The transformed equation (20) can be used for finding a solution of Laplace's equation in a circle, for simplicity the unit circle about the origin, taking on given boundary values $f(\phi)$ on the circumference. The required function $u(r, \phi)$ is to satisfy the equation

(21) $$\frac{\partial^2 u}{\partial r^2} + \frac{1}{r}\ \frac{\partial u}{\partial r} + \frac{1}{r^2}\ \frac{\partial^2 u}{\partial \phi^2} = 0$$

and the condition

$$u(1, \phi) = f(\phi).$$

Tentative assumption of a solution of (21) in the form $u = R(r)F(\phi)$ leads to the equation

$$(r^2 R'' + rR')/R = -F''/F.$$

The quantity thus represented in two different ways must be a constant. If a non-negative value λ^2 is assigned to this constant, the separate equations

$$r^2 R'' + rR' - \lambda^2 R = 0, \qquad F'' + \lambda^2 F = 0$$

are obtained. The latter has the solutions $\cos \lambda\phi$, $\sin \lambda\phi$. Solutions of the former, readily found by standard methods and still more readily verified by substitution, are

r^λ and $r^{-\lambda}$. It may be supposed for definiteness, and without loss of generality, that $\lambda \geqq 0$. To restrict consideration to functions which are continuous at the origin as well as elsewhere, the power of r with a negative exponent may be set aside. In order that different determinations of the polar coordinates of a point, with values of ϕ differing by multiples of 2π, shall lead to the same value of u, integral values are to be assigned to λ. Particular solutions of (21) thus arrived at are

$$r^n \cos n\phi, n = 0, 1, 2, \cdots ; \qquad r^n \sin n\phi, n = 1, 2, \cdots .$$

The fact that these, as real and pure imaginary components of z^n when $z = r(\cos \phi + i \sin \phi)$, are solutions of Laplace's equation, is well known also from the beginnings of the theory of functions of a complex variable.

A solution of the boundary value problem, formally at least, is then

$$(22) \quad u(r, \phi) = \frac{a_0}{2} + \sum_{n=1}^{\infty} r^n(a_n \cos n\phi + b_n \sin n\phi),$$

with coefficients a_n, b_n determined so that

$$f(\phi) = u(1, \phi) = \frac{a_0}{2} + \sum_{n=1}^{\infty} (a_n \cos n\phi + b_n \sin n\phi).$$

The designation $a_0/2$ for the constant term is of course chosen for the sake of agreement with the notation of Chapter I.

The coefficients being represented by the formulas

$$a_n = \frac{1}{\pi} \int_{-\pi}^{\pi} f(\theta) \cos n\theta \, d\theta, \qquad b_n = \frac{1}{\pi} \int_{-\pi}^{\pi} f(\theta) \sin n\theta \, d\theta,$$

the right-hand member of (22) has the form

(23) $\quad \dfrac{1}{\pi} \displaystyle\int_{-\pi}^{\pi} f(\theta) \left[\dfrac{1}{2} + \sum_{n=1}^{\infty} r^n \cos n(\theta - \phi) \right] d\theta.$

Let $\theta - \phi = \alpha$, and let $z = r(\cos \alpha + i \sin \alpha)$. Then $z^n = r^n(\cos n\alpha + i \sin n\alpha)$, and

$$\sum_{n=0}^{\infty} r^n \cos n\alpha$$

is the real part of

$$\sum_{n=0}^{\infty} z^n = \frac{1}{1 - z} = \frac{1}{1 - r \cos \alpha - ir \sin \alpha}$$
$$= \frac{1 - r \cos \alpha + ir \sin \alpha}{1 - 2r \cos \alpha + r^2}.$$

Consequently

$$\sum_{n=0}^{\infty} r^n \cos n\alpha = \frac{1 - r \cos \alpha}{1 - 2r \cos \alpha + r^2},$$
$$\frac{1}{2} + \sum_{n=1}^{\infty} r^n \cos n\alpha = \frac{1 - r \cos \alpha}{1 - 2r \cos \alpha + r^2} - \frac{1}{2}$$
$$= \frac{1}{2} \cdot \frac{1 - r^2}{1 - 2r \cos \alpha + r^2}.$$

Substitution of this in (23) gives

(24) $\quad u(r, \phi) = \dfrac{1}{2\pi} \displaystyle\int_{-\pi}^{\pi} f(\theta) \, \dfrac{1 - r^2}{1 - 2r \cos (\theta - \phi) + r^2} \, d\theta.$

The expression (24) for the solution of the boundary value problem is known as *Poisson's integral*.

7. Transformation of Laplace's equation in three dimensions. Attention is to be turned now to the three-dimensional equation (1). Direct treatment of this equation in rectangular coordinates is reserved for the next chapter. The present concern is with its transformation into cylindrical and spherical coordinates.

The variables (r, ϕ, ζ) or (r, ϕ, z) are *cylindrical coordinates* if they are related to the rectangular coordinates (x, y, z) by the equations

$$x = r \cos \phi, \qquad y = r \sin \phi, \qquad z = \zeta.$$

Since the third coordinate is the same in both systems it will be represented by the same letter z, without further use of the alternative ζ. The r-coordinate represents distance from the z-axis. The pair (r, ϕ) can be regarded as polar coordinates of the projection of the point (x, y, z) on the (x, y) plane. The derivatives of a function u with respect to x and y are transformed exactly as in §5. The derivative $\partial^2 u/\partial z^2$ undergoes no change. Consequently

$$(25) \quad \frac{\partial^2 u}{\partial x^2} + \frac{\partial^2 u}{\partial y^2} + \frac{\partial^2 u}{\partial z^2} = \frac{\partial^2 u}{\partial r^2} + \frac{1}{r} \frac{\partial u}{\partial r} + \frac{1}{r^2} \frac{\partial^2 u}{\partial \phi^2} + \frac{\partial^2 u}{\partial z^2}.$$

A system of *spherical coordinates* is defined by the relations

$$x = \rho \sin \theta \cos \phi,$$
$$y = \rho \sin \theta \sin \phi,$$
$$z = \rho \cos \theta.$$

Here θ may be regarded as colatitude and ϕ as longitude, the latter coordinate having the same geometric meaning as in the preceding paragraph; ρ is distance from the origin. If the derivatives of a function u with respect to ρ, θ, ϕ are calculated in terms of $\partial u/\partial x, \partial u/\partial y, \partial u/\partial z$, and the resulting equations solved for the latter derivatives, it is found that

$$\frac{\partial u}{\partial x} = \sin\theta\cos\phi\,\frac{\partial u}{\partial\rho} + \frac{1}{\rho}\cos\theta\cos\phi\,\frac{\partial u}{\partial\theta} - \frac{\sin\phi}{\rho\sin\theta}\,\frac{\partial u}{\partial\phi},$$

$$\frac{\partial u}{\partial y} = \sin\theta\sin\phi\,\frac{\partial u}{\partial\rho} + \frac{1}{\rho}\cos\theta\sin\phi\,\frac{\partial u}{\partial\theta} + \frac{\cos\phi}{\rho\sin\theta}\,\frac{\partial u}{\partial\phi},$$

$$\frac{\partial u}{\partial z} = \cos\theta\,\frac{\partial u}{\partial\rho} - \frac{1}{\rho}\sin\theta\,\frac{\partial u}{\partial\theta}.$$

Further substitution for the evaluation of the second derivatives, in analogy with the procedure followed in §5, is laborious, but not prohibitively so. It can be shown by straightforward calculation that

$$(26) \quad \begin{aligned} \frac{\partial^2 u}{\partial x^2} + \frac{\partial^2 u}{\partial y^2} + \frac{\partial^2 u}{\partial z^2} &= \frac{1}{\rho^2}\frac{\partial}{\partial\rho}\left(\rho^2\frac{\partial u}{\partial\rho}\right) \\ &+ \frac{1}{\rho^2\sin\theta}\frac{\partial}{\partial\theta}\left(\sin\theta\frac{\partial u}{\partial\theta}\right) + \frac{1}{\rho^2\sin^2\theta}\frac{\partial^2 u}{\partial\phi^2}. \end{aligned}$$

An alternative derivation based on Green's theorem is very much more expeditious, but less automatic in its operation.

8. Legendre series: Laplace's equation in a sphere. Spherical coordinates may be used in finding a solution of Laplace's equation which takes on given values on the surface of a sphere. The sphere will be taken as of unit radius, with center at the origin. The problem will be considered for the present only in the simplified form which results if the boundary values are supposed to be independent of the longitude coordinate ϕ. It may then plausibly be assumed, and is in fact necessarily true, that the solution of the problem also is independent of ϕ. This implies that the last term of (26) is absent, the derivative $\partial^2 u/\partial\phi^2$ being identically zero. A function $u(\rho, \theta)$ is to be found satisfying the differential equation

$$(27) \qquad \frac{\partial}{\partial \rho}\left(\rho^2 \frac{\partial u}{\partial \rho}\right) + \frac{1}{\sin \theta} \frac{\partial}{\partial \theta}\left(\sin \theta \frac{\partial u}{\partial \theta}\right) = 0$$

with the boundary condition

$$(28) \qquad\qquad u(1, \theta) = f(\theta), \qquad\qquad 0 \leqq \theta \leqq \pi,$$

the function $f(\theta)$ being given. The more general problem in which the variable ϕ is permitted to appear will be discussed in the next chapter.

If a solution of (27) is assumed in the form $u = R(\rho)T(\theta)$, the equation which R and T must satisfy is

$$\frac{\rho^2 R'' + 2\rho R'}{R} = -\frac{T'' + \cot \theta\, T'}{T}.$$

The common value of these two expressions must be a constant. A sufficiency of particular solutions will be obtained by taking the constant as non-negative. For convenience in the subsequent calculation let it be represented by $n(n+1)$, $n \geqq 0$. The equation

$$\rho^2 R'' + 2\rho R' - n(n + 1)R = 0$$

has ρ^n and ρ^{-n-1} as solutions. In the relation

$$(29) \qquad \frac{d^2 T}{d\theta^2} + \cot \theta\, \frac{dT}{d\theta} + n(n + 1)T = 0$$

let a new independent variable be introduced by the substitution $x = \cos \theta$. (This x is of course not the same as the x of the rectangular coordinate system.) Then

$$\frac{dT}{d\theta} = -\sin \theta\, \frac{dT}{dx},$$

$$\frac{d^2 T}{d\theta^2} = \sin^2 \theta\, \frac{d^2 T}{dx^2} - \cos \theta\, \frac{dT}{dx} = (1 - x^2)\frac{d^2 T}{dx^2} - x\, \frac{dT}{dx},$$

and (29) becomes

$$(1 - x^2) \frac{d^2T}{dx^2} - 2x \frac{dT}{dx} + n(n + 1)T = 0.$$

This is the differential equation of the Legendre polynomials. For non-negative integral n it has a solution $P_n(x) = P_n(\cos \theta)$ which is continuous for all values of x, and in particular for $-1 \leqq x \leqq 1$ and so for all real values of θ. For the sake of continuity at the origin the value ρ^n is to be chosen for R in preference to ρ^{-n-1}. The corresponding particular solutions of (27) are the functions $\rho^n P_n(\cos \theta)$, $n = 0, 1, 2, \cdots$.

To satisfy the boundary condition (28) it remains to set

$$u(\rho, \theta) = \sum_{n=0}^{\infty} a_n \rho^n P_n(\cos \theta),$$

$$f(\theta) = u(1, \theta) = \sum_{n=0}^{\infty} a_n P_n(\cos \theta).$$

If $f(\theta)$ as a function of $\cos \theta$ is denoted by $F(x)$, the coefficients a_n are to be determined by expanding $F(x)$ in a Legendre series $\sum a_n P_n(x)$.

9. Bessel series: Laplace's equation in a cylinder. Let the boundary surfaces of a cylindrical volume be in cylindrical coordinates the planes $z = 0$ and $z = b$ and the surface $r = 1$. A boundary value problem leading to an expansion in Bessel series may be formulated by requiring a solution of Laplace's equation which takes on given values on the lower base of this cylinder, and vanishes on the upper base and on the curved surface. The problem will be simplified once more by assuming that the functions involved are independent of the ϕ-coordinate, the more general problem being reserved for later treatment. Then the question is that of finding a function $u(r, z)$ which, in view of the vanishing of the next to the last term in (25), satisfies the differential equation

$$(30) \qquad \frac{\partial^2 u}{\partial r^2} + \frac{1}{r} \frac{\partial u}{\partial r} + \frac{\partial^2 u}{\partial z^2} = 0,$$

with the auxiliary conditions

$$(31) \qquad u(1, z) = 0, \qquad u(r, b) = 0, \qquad u(r, 0) = f(r)$$

for a given function $f(r)$.

Substitution of a product $R(r) Z(z)$ for u in (30) leads to the equations

$$(32) \qquad \frac{R'' + (R'/r)}{R} = - \frac{Z''}{Z} = - \lambda^2,$$

if the constant value of the first two members is supposed to be negative or zero. The equation $Z'' = \lambda^2 Z$ has for solutions $e^{\lambda z}$ and $e^{-\lambda z}$, and as a combination of these vanishing for $z = b$ the function $z = \sinh \lambda(b - z)$. For R the equation

$$R'' + \frac{1}{r} R' + \lambda^2 R = 0$$

has by the first paragraph of §3 in Chapter III the solution $J_0(\lambda r)$. If λ is one of the numbers λ_n for which $J_0(\lambda_n) = 0$, the product

$$\sinh \lambda_n(b - z) J_0(\lambda_n r)$$

satisfies all the requirements (30), (31) with the exception of that relating to the arbitrary function $f(r)$. To meet this condition also let

$$u(r, z) = \sum_{n=1}^{\infty} c_n \sinh \lambda_n(b - z) J_0(\lambda_n r),$$

$$f(r) = \sum_{n=1}^{\infty} a_n J_0(\lambda_n r), \qquad c_n = a_n / \sinh \lambda_n b.$$

The solution of the boundary value problem is accomplished by expanding $f(r)$ in a Bessel series of the type indicated.

Slight changes of detail would adapt the formulas to arbitrarily assigned values (independent of ϕ) on the upper base of the cylinder, with zero values on the curved surface and on the lower base.

It is appropriate in this connection to note also the form of the solution when arbitrary values independent of ϕ are assigned on the curved surface, with vanishing boundary values on both bases. In symbols, the boundary values are of the form

$$u(r, 0) = 0, \qquad u(r, b) = 0, \qquad u(1, z) = F(z).$$

It is convenient now to replace $-\lambda^2$ in (32) by λ^2. The function $Z = \sin \lambda z$ satisfies the equation $Z'' = -\lambda^2 Z$ and vanishes for $z = 0$, and vanishes also for $z = b$ if λ has the form $n\pi/b$ with an integral value of n. The equation

$$R'' + \frac{1}{r} R' - \lambda^2 R = 0$$

has in terms of the notation used hitherto the solution $J_0(i\lambda r)$, as appears formally on replacement of λ above by $i\lambda$; the function $J_0(ix)$ is however no more imaginary than $J_0(x)$, being represented by the series

$$1 + \frac{x^2}{2^2} + \frac{x^4}{2^2 4^2} + \frac{x^6}{2^2 4^2 6^2} + \cdots$$

with positive coefficients throughout. If the function defined by this series is denoted by $I_0(x)$, the form obtained for R is $I_0(\lambda r)$. The differential equation and the first two boundary conditions have the solutions

$$I_0(n\pi r/b) \sin (n\pi z/b), \qquad n = 1, 2, 3, \cdots.$$

The function $u(r, z)$ is then to be represented by a series of these functions, with coefficients determined by expanding $F(z)$ in a sine series.

By combination of the solutions that have been described the boundary value problem can be solved for arbitrarily given boundary values independent of ϕ on the entire surface of the cylinder.

If the first condition in (31) is replaced by $u_r(1, z) = 0$, the solution proceeds as before except that roots of the equation $J_0'(x) = 0$ are to be used instead of the roots of $J_0(x) = 0$. This condition, as well as the original one and another alternative calling for the third set of roots described in §3 of Chapter III, has a simple physical interpretation.*

10. Bessel series: circular drumhead. The vibrations of a stretched elastic membrane, under the simplest conditions of uniformity and freedom from retarding forces, are described by the equation

$$\frac{\partial^2 z}{\partial t^2} = a^2 \left(\frac{\partial^2 z}{\partial x^2} + \frac{\partial^2 z}{\partial y^2} \right),$$

where (x, y) are coordinates in the plane of the membrane in its position of equilibrium, and z measures displacement perpendicular to that plane. Let such a membrane in particular be fixed at all points of the circumference of unit radius about the origin. If (r, ϕ) denote

---

* See e.g. Byerly, p. 227.

polar coordinates in the plane, these together with z constitute a set of cylindrical coordinates in space. Let the membrane be given an initial displacement independent of ϕ, i.e. having a value $f(r)$ at all points at distance r from the center, and released from rest. Then the displacement at any subsequent time, assumed to be independent of ϕ throughout, is represented by a function $z(r, t)$ satisfying the equations

$$(33) \qquad \frac{\partial^2 z}{\partial t^2} = a^2\left(\frac{\partial^2 z}{\partial r^2} + \frac{1}{r}\,\frac{\partial z}{\partial r}\right),$$

$$(34) \qquad z(1, t) = 0, \qquad z(r, 0) = f(r), \qquad z_t(r, 0) = 0.$$

Particular solutions of (33) come from the relations $z(r, t) = R(r)\ T(t)$,

$$\frac{T''}{a^2 T} = \frac{R'' + (R'/r)}{R} = -\lambda^2,$$

$$T = \cos \lambda at, \ \sin \lambda at, \qquad R = J_0(\lambda r).$$

The first and third of (34) are satisfied by $J_0(\lambda_n r) \cos \lambda_n at$, if λ_n is a root of $J_0(x) = 0$. The remaining condition gives rise to the formulas

$$z(r, t) = \sum_{n=1}^{\infty} a_n J_0(\lambda_n r) \cos \lambda_n at,$$

$$f(r) = z(r, 0) = \sum_{n=1}^{\infty} a_n J_0(\lambda_n r).$$

If $f(r)$ has the special form of a single term $hJ_0(\lambda_k r)$, z reduces to $hJ_0(\lambda_k r) \cos \lambda_k at$, and any one point of the membrane, corresponding to a fixed r, vibrates in a simple harmonic motion with period $2\pi/(\lambda_k a)$. But when a number of such vibrations are superposed their frequencies are not simply related like those of the harmonics of an undamped vibrating string, and the sound given out by a vibrating drumhead is discordant even without the effect of resisting forces.

If the drumhead, more practically, is set in vibration by a blow, instead of being released from a distorted position, the process of solution is similar, involving sine factors in place of cosines.

The examples treated so far have involved the Bessel functions $J_n(x)$ only for $n = 0$. The next chapter will give occasion for the use of more general Bessel functions.

Supplementary references: Byerly; Churchill; Kellogg; Courant-Hilbert; Riemann-Weber.

CHAPTER V

DOUBLE SERIES; LAPLACE SERIES

1. Boundary value problem in a cube; double Fourier series. More general boundary value problems than those of the last chapter call for the expansion of functions of two independent variables in series of special functions which are orthogonal with respect to integration over a two-dimensional domain. One type of such series, called Laplace series, is of highly distinctive character, and will be studied in some detail below. An example will be presented first which leads to a form of series with less striking features of novelty.

Let a function $f(x, y)$ be defined for $0 \leqq x \leqq \pi$, $0 \leqq y \leqq \pi$, and let a boundary value problem be formulated for a cube of corresponding dimensions by requiring a function $u(x, y, z)$ which satisfies Laplace's equation

$$(1) \qquad \frac{\partial^2 u}{\partial x^2} + \frac{\partial^2 u}{\partial y^2} + \frac{\partial^2 u}{\partial z^2} = 0$$

with the auxiliary conditions

$$u(x, y, 0) = f(x, y), \qquad u(x, y, \pi) = 0,$$
$$u(0, y, z) = 0, \qquad u(\pi, y, z) = 0,$$
$$u(x, 0, z) = 0, \qquad u(x, \pi, z) = 0.$$

The result of substituting a product $X(x)\ Y(y)\ Z(z)$ for u in (1) can be written in the form

$$-\frac{X''}{X} = \frac{Y''}{Y} + \frac{Z''}{Z}.$$

By a type of argument used repeatedly in the preceding chapter, the quantity thus represented in two ways, being on the one hand independent of y and z and on the other hand independent of x, is a constant, which for the purposes of the present problem may be taken as non-negative and denoted by λ^2. Then

$$-\frac{Y''}{Y} = \frac{Z''}{Z} - \lambda^2.$$

The common value of the left and right members here is again a constant, which will be called μ^2. The equations

$$X'' = -\lambda^2 X, \qquad Y'' = -\mu^2 Y, \qquad Z'' = (\lambda^2 + \mu^2)Z$$

have the solutions

$$\cos \lambda x, \ \sin \lambda x; \ \cos \mu y, \ \sin \mu y; \ e^{\nu z}, \ e^{-\nu z}, \ \nu = (\lambda^2 + \mu^2)^{1/2}.$$

A linear combination of the last pair is

$$\sinh (\lambda^2 + \mu^2)^{1/2}(\pi - z),$$

which vanishes for $z = \pi$. If λ, μ are given integral values m, n, a function satisfying (1) and five of the six boundary conditions is seen to be

$$\sin mx \sin ny \sinh (m^2 + n^2)^{1/2}(\pi - z).$$

The remaining condition is satisfied by

$$u(x, y, z)$$

$$(2) \qquad = \sum_{m=1}^{\infty} \sum_{n=1}^{\infty} h_{mn} \sin mx \sin ny \sinh (m^2 + n^2)^{1/2}(\pi - z)$$

if $f(x, y)$ can be expanded in a series of the form

$$(3) \qquad \sum_{m=1}^{\infty} \sum_{n=1}^{\infty} d_{mn} \sin mx \sin ny,$$

the coefficient h_{mn} in (2) being $d_{mn}/\sinh (m^2+n^2)^{1/2}\pi$. The series (3) is a *double Fourier series*, of particularly simple form to be sure, inasmuch as it involves only sine terms.

Any two of the functions $\sin mx \sin ny$, $\sin px \sin qy$ are orthogonal to each other for integration with respect to x and y over the square $0 \leqq x \leqq \pi$, $0 \leqq y \leqq \pi$, i.e. the integral of their product is zero, unless $m = p$ and $n = q$; for the double integral of the product over the square is merely the product of the integrals

$$\int_0^\pi \sin mx \sin px\, dx, \qquad \int_0^\pi \sin ny \sin qy\, dy.$$

When $m = p$, $n = q$ the product is not zero but $\pi^2/4$. By the usual procedure with series of orthogonal functions, the coefficients in (3) are determined formally to be

$$d_{mn} = \frac{4}{\pi^2} \int_0^\pi \int_0^\pi f(x,\, y) \sin mx \sin ny\, dx\, dy.$$

Double Fourier series of greater generality involve terms of the four types

$$\cos mx \cos ny, \quad \cos mx \sin ny, \quad \sin mx \cos ny, \quad \sin mx \sin ny.$$

The formulas are further complicated by the necessity of special attention to the terms containing cosine factors of order zero in one variable or both. The detailed study of double Fourier series must be omitted here.

With regard to the boundary value problem for the cube it is clear that what has been done for the lower base could be adapted to any of the other five faces. By combination of the results a formal solution would be obtained for arbitrary boundary values on the entire surface of the cube. The formulas could be further generalized at once to a rectangular parallelepiped of arbitrary dimensions.

2. General spherical harmonics. The construction of particular solutions of the general Laplace equation in spherical coordinates is a matter calling for examination at some length.

The equation, by (26) of Chapter IV, is

$$(4) \quad \frac{\partial}{\partial \rho}\left(\rho^2 \frac{\partial u}{\partial \rho}\right) + \frac{1}{\sin \theta} \frac{\partial}{\partial \theta}\left(\sin \theta \frac{\partial u}{\partial \theta}\right) + \frac{1}{\sin^2 \theta} \frac{\partial^2 u}{\partial \phi^2} = 0.$$

By substitution of $R(\rho) \; T(\theta) \; F(\phi)$ for u this takes the form

$$(5) \quad \frac{\rho^2 R'' + 2\rho R'}{R} + \frac{T'' + \cot \theta \, T'}{T} + \frac{1}{\sin^2 \theta} \frac{F''}{F} = 0.$$

The first fraction, independent of θ and ϕ as it stands and independent of ρ by virtue of the equation, is a constant; let this be denoted by $m(m+1)$. On multiplication of (5) by $\sin^2 \theta$ it appears that F''/F is independent of ϕ as well as independent of ρ and θ; let it be set equal to $-n^2$. Thus R and F satisfy the equations

$$(6) \qquad \rho^2 R'' + 2\rho R' - m(m+1)R = 0,$$

$$(7) \qquad \qquad F'' + n^2 F = 0.$$

When the first fraction and F''/F are replaced by their constant values, (5) becomes

$$(8) \quad T'' + \cot \theta \, T' + \left[m(m+1) - \frac{n^2}{\sin^2 \theta} \right] T = 0.$$

Equation (6), as in §8 of Chapter IV (except for a slight difference of notation), has the solutions ρ^m and ρ^{-m-1}. Of these the former will be preferred once more, as continuous for $\rho = 0$, m being supposed non-negative. Solutions of (7) are $\cos n\phi$ and $\sin n\phi$. Here n will be taken as non-negative and integral. The purpose of the restriction to integral values is to avoid the occurrence of multiple values of u at a point of space in consequence of alternative determinations of the longitude coordinate, differing by integral multiples of 2π.

The equation (8) differs from (29) of Chapter IV, apart from notation, only by the presence of the term $-n^2/\sin^2 \theta$ inside the brackets. By the substitution $x = \cos \theta$ this becomes $-n^2/(1-x^2)$, and when the substitution is carried through in the other terms, as in Chapter IV, the whole equation takes the form

$$(9) \quad (1-x^2) \frac{d^2T}{dx^2} - 2x \frac{dT}{dx} + \left[m(m+1) - \frac{n^2}{1-x^2} \right] T = 0.$$

A procedure for arriving at the desired form of solution of this equation, while perhaps not obvious *a priori*, is readily verified when the essential steps are suggested. Let a new dependent variable be introduced by setting $T = (1-x^2)^{n/2}z$ (not the z of the rectangular coordinate system). Substitution of this expression for T in (9), with some combination of terms and division of a factor $(1-x^2)^{n/2}$ from the whole left-hand member, leads to the equation

$$(10) \quad (1-x^2)z'' - 2(n+1)xz' + [m(m+1) - n(n+1)]z = 0.$$

On the other hand, if $y = P_m(x)$, repeated differentiation of

$$(1 - x^2)y'' - 2xy' + m(m + 1)y = 0$$

gives

$$(1-x^2)y''' - 4xy'' + [m(m+1) - 2]y' = 0,$$
$$(1-x^2)y^{iv} - 6xy''' + [m(m+1) - 6]y'' = 0,$$

. . . .

$$(1-x^2)y^{(n+2)} - 2(n+1)xy^{(n+1)} + [m(m+1) - n(n+1)]y^{(n)} = 0,$$

which means that the function $z = d^n y/dx^n$ satisfies (10). Consequently (9) is satisfied for non-negative integral m and n by

$$T = (1 - x^2)^{n/2} \frac{d^n}{dx^n} P_m(x),$$

which in the notation of (8) is

$$T = \sin^n \theta \frac{d^n}{(d \cos \theta)^n} P_m(\cos \theta).$$

As this is identically zero for $n > m$, the result is nontrivial only for $n = 0, 1, \cdots, m$.

With the notation $P_m^{(n)}(x)$ or $P_m^{(n)}(\cos \theta)$ for the derivatives in the last expressions, let

$$u_{mn} = \cos n\phi \sin^n \theta P_m^{(n)}(\cos \theta), \qquad n = 0, 1, 2, \cdots, m,$$

$$v_{mn} = \sin n\phi \sin^n \theta P_m^{(n)}(\cos \theta), \qquad n = 1, 2, \cdots, m.$$

Particular solutions obtained for (4) are $\rho^m u_{mn}$ and $\rho^m v_{mn}$. These functions of (ρ, θ, ϕ) are called *spherical harmonics*. The same designation is also applied to u_{mn} and v_{mn} without the factor ρ^m, as functions of θ and ϕ.

For each m, the $2m+1$ functions u_{mn}, v_{mn} (or the corresponding functions with the common factor ρ^m included) are linearly independent; that is to say, there is no set of constants $A_0, A_1, \cdots, A_m, B_1, \cdots, B_m$, not all zero, such that

$$(11) \qquad A_0 u_{m0} + \sum_{n=1}^{m} (A_n u_{mn} + B_n v_{mn})$$

vanishes identically. For an expression

$$(12) \qquad \alpha_0 + \sum_{n=1}^{m} (\alpha_n \cos n\phi + \beta_n \sin n\phi)$$

can not be identically zero unless the coefficients $\alpha_0, \cdots, \alpha_m, \beta_1, \cdots, \beta_m$ are all zero; the sum (11) has the form (12) with $\alpha_0 = A_0 P_m(\cos \theta)$,

$$\alpha_n = A_n \sin^n \theta P_m^{(n)}(\cos \theta), \qquad \beta_n = B_n \sin^n \theta P_m^{(n)}(\cos \theta),$$
$$n = 1, 2, \cdots, m,$$

and its identical vanishing in θ and ϕ would require that $\alpha_0, \alpha_1, \cdots, \alpha_m, \beta_1, \cdots, \beta_m$ vanish for each value of θ, i.e., vanish identically in θ, which is not the case unless the A's and B's are all zero. The fact of linear independence is perhaps obvious, but is emphasized because it will be needed explicitly later.

3. Laplace series. By means of the functions of the preceding section the solution of the boundary value problem for Laplace's equation in a sphere (see Chapter IV, §8) can be extended to values which are not independent of ϕ. A function $u(\rho, \theta, \phi)$ is to satisfy (4) together with a boundary condition of the form

$$u(1, \theta, \phi) = f(\theta, \phi).$$

To yield such a function, in analogy with the procedure followed in other cases after suitable particular solutions of the differential equation have been obtained, the coefficients in a representation

$$u(\rho, \theta, \phi) = \sum_{m=0}^{\infty} \rho^m \left[\frac{1}{2} a_{m0} u_{m0} + \sum_{n=1}^{m} (a_{mn} u_{mn} + b_{mn} v_{mn}) \right]$$

are determined so that

$$(13) \quad \begin{aligned} f(\theta, \phi) &= u(1, \theta, \phi) \\ &= \sum_{m=0}^{\infty} \left[\frac{1}{2} a_{m0} u_{m0} + \sum_{n=1}^{m} (a_{mn} u_{mn} + b_{mn} v_{mn}) \right]. \end{aligned}$$

(For $m = 0$ it is understood that the expression in brackets reduces to the single term $\frac{1}{2}a_{00}$.) A series of the form (13) is called a *Laplace series*.

In the Laplace series, as in the other types that have been considered, the determination of the coefficients is facilitated by properties of orthogonality of the functions which make up the terms. On the surface of the sphere, θ varies from 0 to π, and ϕ may be taken as varying from $-\pi$ to π. The element of area is $\sin\theta \, d\phi \, d\theta$. The integral of a function $F(\theta, \phi)$ *over the surface of the sphere* means

$$\int_0^\pi \int_{-\pi}^\pi F(\theta, \phi) \sin\theta \, d\phi \, d\theta.$$

Two functions are orthogonal to each other if the integral of their product is zero. *Any two of the functions u_{mn}, u_{rs} are orthogonal to each other for integration over the surface of the sphere unless $m = r$ and $n = s$; any two distinct v's are similarly orthogonal; and any u_{mn} is orthogonal to any v_{rs}, whether the subscripts are the same or different.*

The integral of any one of the products concerned reduces at once to a product of two simple integrals. For example, the integral of $u_{mn}u_{rs}$ over the sphere is

$$\int_{-\pi}^\pi \cos n\phi \, \cos s\phi \, d\phi \int_0^\pi \sin^{n+s+1}\theta \, P_m^{(n)}(\cos\theta) P_r^{(s)}(\cos\theta) d\theta.$$

The first factor is zero unless $n = s$. Similarly, the orthogonality of v_{mn} and v_{rs} for $n \neq s$, and of u_{mn} and v_{rs} without restriction on the subscripts, is an immediate consequence of the elementary properties of the trigonometric functions. For $n = s$ it remains to examine the integral in terms of θ, which then becomes

$$(14) \qquad \int_0^\pi \sin^{2n+1}\theta \, P_m^{(n)}(\cos\theta) P_r^{(n)}(\cos\theta) d\theta,$$

being the same whether a product of u's or a product of v's is under consideration.

In terms of the variable $x = \cos\theta$ (quite distinct once more from the rectangular x-coordinate) (14) is equal to

$$(15) \qquad \int_{-1}^1 (1-x^2)^n P_m^{(n)}(x) P_r^{(n)}(x) dx.$$

Since $P_m^{(n)}(x)$ is a solution of (10),

$$(1-x^2)\frac{d^2}{dx^2} P_m^{(n)}(x) - 2(n+1)x\frac{d}{dx} P_m^{(n)}(x)$$
$$+ [m(m+1) - n(n+1)]P_m^{(n)}(x) = 0.$$

Let this equation, multiplied by $(1-x^2)^n P_r^{(n)}(x)$, be subtracted from the corresponding equation with m and r interchanged. The index n is the same throughout. With

$$w = P_r^{(n+1)}(x)P_m^{(n)}(x) - P_m^{(n+1)}(x)P_r^{(n)}(x),$$

the result can be arranged in the form

$$(16) \quad \begin{aligned} &(1-x^2)^{n+1}\frac{dw}{dx} - 2(n+1)x(1-x^2)^n w \\ &= [m(m+1) - r(r+1)](1-x^2)^n P_m^{(n)}(x) P_r^{(n)}(x). \end{aligned}$$

The left member is the derivative of $(1-x^2)^{n+1}w$; its definite integral from -1 to 1 vanishes. The quantity in brackets on the right is different from zero if r and m are non-negative and distinct. So integration of (16) from -1 to 1 shows that

$$\int_{-1}^{1} (1 - x^2)^n P_m^{(n)}(x) P_r^{(n)}(x) dx = 0.$$

It is of importance also to find the value of (15) when $m = r$. For $n = 0$ it is known from Chapter II that

$$\int_{-1}^{1} [P_m(x)]^2 dx = \frac{2}{2m + 1}.$$

For integration by parts over the interval $(-1, 1)$, let the integral

$$\int (1 - x^2) [P_m'(x)]^2 dx$$

be regarded as having the form $\int u dv$, with

$$u = (1 - x^2) P_m'(x), \qquad v = P_m(x).$$

In consequence of the differential equation which $P_m(x)$ satisfies,

$$du = [(1 - x^2) P_m''(x) - 2x P_m'(x)] dx = -m(m+1) P_m(x) dx.$$

The product uv vanishes for $x = \pm 1$, and therefore

$$\int_{-1}^{1} (1 - x^2) [P_m'(x)]^2 dx = m(m + 1) \int_{-1}^{1} [P_m(x)]^2 dx$$

$$= \frac{2m(m + 1)}{2m + 1}.$$

For general n let a similar process of integration by parts be applied in

$$\int_{-1}^{1} (1 - x^2)^{n+1} \left[P_m^{(n+1)}(x) \right]^2 dx$$

with

$$u = (1 - x^2)^{n+1} P_m^{(n+1)}(x), \ dv = P_m^{(n+1)}(x)dx, \ v = P_m^{(n)}(x).$$

Since $P_m^{(n)}(x)$ satisfies the differential equation (10),

$$du = \left[(1-x^2)^{n+1} P_m^{(n+2)}(x) - 2(n+1)x(1-x^2)^n P_m^{(n+1)}(x)\right]dx$$

$$= -\left[m(m+1) - n(n+1)\right](1-x^2)^n P_m^{(n)}(x)dx$$

$$= -(m-n)(m+n+1)(1-x^2)^n P_m^{(n)}(x)dx.$$

It is still true that uv vanishes at both ends of the interval. So

$$\int_{-1}^{1} (1 - x^2)^{n+1} \left[P_m^{(n+1)}(x)\right]^2 dx$$

$$= (m - n)(m + n + 1) \int_{-1}^{1} (1 - x^2)^n \left[P_m^{(n)}(x)\right]^2 dx.$$

This relation makes it possible to calculate the value of the integral for successive values of n by induction. It is found that

$$\int_{-1}^{1} (1 - x^2)^n \left[P_m^{(n)}(x)\right]^2 dx$$

$$= (m - n + 1)(m - n + 2) \cdots (m + n) \frac{2}{2m + 1}$$

$$= \frac{(m + n)!}{(m - n)!} \frac{2}{2m + 1}.$$

For θ as variable of integration this is then the value of (14) with $m = r$.

The usual process for finding the coefficients in a series of orthogonal functions now gives for a_{mn} and b_{mn} in (13) the representations

$$a_{mn} = \frac{(m-n)!}{(m+n)!}\frac{2m+1}{2\pi}\times$$

$$\int_0^\pi \int_{-\pi}^\pi f(\theta,\phi)\sin^{n+1}\theta\cos n\phi P_m^{(n)}(\cos\theta)d\phi\,d\theta,$$

(17)

$$b_{mn} = \frac{(m-n)!}{(m+n)!}\frac{2m+1}{2\pi}\times$$

$$\int_0^\pi \int_{-\pi}^\pi f(\theta,\phi)\sin^{n+1}\theta\sin n\phi P_m^{(n)}(\cos\theta)d\phi\,d\theta.$$

The formula for a_{mn} holds for $n=0$ as well as for positive n.

4. Harmonic polynomials.* It is to be shown next that each of the spherical harmonics $\rho^m u_{mn}$, $\rho^m v_{mn}$ when considered as a function of the rectangular coordinates (x, y, z) is a homogeneous polynomial of the mth degree, i.e. one in which each term is of the mth degree in the three variables together. A function which satisfies Laplace's equation is called a *harmonic function*. Each of the functions $\rho^m u_{mn}$, $\rho^m v_{mn}$ is a homogeneous harmonic polynomial.

It is possible to represent $\cos n\phi$ in a variety of ways as a polynomial in $\cos\phi$ and $\sin\phi$. In particular, by de Moivre's theorem, i.e. by separation of real and pure imaginary parts in

$$\cos n\phi + i\sin n\phi = (\cos\phi + i\sin\phi)^n,$$

it can be expressed as a sum of terms of the form

---

* This section and the next three deal with questions which are important in connection with the study of the Laplace series for its own sake, but can be omitted as far as an understanding of the rest of the book is concerned.

$C \cos^{n-k} \phi \sin^k \phi$, in which C is a constant coefficient, differing from term to term, and k is a non-negative even integer. There is a corresponding representation of $\sin n\phi$ in which each value of k is odd. The terms of $P_m^{(n)}(\cos \theta)$ as a polynomial in $\cos \theta$ are constant multiples of $\cos^{m-n-2h} \theta$ for non-negative integral values of h. So each term of $\rho^m u_{mn}$ or $\rho^m v_{mn}$ is of the form

$$C\rho^m \cos^{n-k} \phi \sin^k \phi \sin^n \theta \cos^{m-n-2h} \theta$$
$$= C\rho^{2h} \cdot \rho^{n-k} \sin^{n-k} \theta \cos^{n-k} \phi \cdot \rho^k \sin^k \theta \sin^k \phi \cdot$$
$$\rho^{m-n-2h} \cos^{m-n-2h} \theta$$
$$= C(x^2 + y^2 + z^2)^h x^{n-k} y^k z^{m-n-2h},$$

and this is a polynomial in (x, y, z), each of whose terms is of degree m.

To set up from first principles an enumeration of all possible homogeneous harmonic polynomials of the mth degree, let $U(x, y, z)$ denote any such polynomial, let the products $x^i y^j z^k$ in which $i+j+k=m$ be arranged in a triangular schedule in the form

$$
\begin{array}{ccccccc}
x^m & x^{m-1}y & x^{m-2}y^2 & \cdots & x^2 y^{m-2} & xy^{m-1} & y^m \\
x^{m-1}z & x^{m-2}yz & x^{m-3}y^2z & \cdots & xy^{m-2}z & y^{m-1}z & \\
x^{m-2}z^2 & x^{m-3}yz^2 & x^{m-4}y^2z^2 & \cdots & y^{m-2}z^2 & & \\
\cdot & \cdot & \cdot & \cdot & \cdot & \cdot & \\
\cdot & \cdot & \cdot & \cdot & \cdot & \cdot & \\
xz^{m-1} & yz^{m-1} & & & & & \\
z^m & & & & & & \\
\end{array}
$$

(18)

and let the coefficient of $x^i y^j z^k$ in U be c_{ijk}. Let the coefficients c_{ijk} be thought of as written in a corresponding triangular array. Let it be supposed that $m \geqq 2$; all polynomials of degrees 0 and 1 are harmonic.

The function

$$(19) \qquad \frac{\partial^2 U}{\partial x^2} + \frac{\partial^2 U}{\partial y^2} + \frac{\partial^2 U}{\partial z^2}$$

is a sum of terms of the form $D_{\alpha\beta\gamma} x^\alpha y^\beta z^\gamma$ in which $\alpha+\beta+\gamma = m-2$. The condition that U be harmonic is that each of the coefficients $D_{\alpha\beta}$ vanish. The value of the general coefficient $D_{\alpha\beta\gamma}$ is

$$(\alpha + 2)(\alpha + 1)c_{\alpha+2,\beta,\gamma} + (\beta + 2)(\beta + 1)c_{\alpha,\beta+2,\gamma}$$
$$+ (\gamma + 2)(\gamma + 1)c_{\alpha,\beta,\gamma+2}.$$

Setting this expression equal to zero determines $c_{\alpha,\ \beta,\ \gamma+2}$, for example, if the other two c's are known. In the triangular array of all the c's the row in which any coefficient stands is indicated by its third subscript, which is 0 for the first row, 1 for the second, and so on; $c_{\alpha,\ \beta,\ \gamma+2}$ is in a row not earlier than the third, while $c_{\alpha+2,\ \beta,\ \gamma}$ and $c_{\alpha,\ \beta+2,\ \gamma}$ are in the second row above it. If the c's in the first row are given, those in the third row are thereby determined, these determine the fifth row, and so on. The second row determines the succeeding even-numbered rows. If arbitrary values are assigned to the first two rows of c's, the rest can be found so as to make (19) vanish identically.

For example, there is one and just one homogeneous harmonic polynomial in which x^m has coefficient unity and all the other monomials from the first two rows of (18) have zero coefficients. There is one and just one in which $x^{m-1}y$ is the only term from these two rows, and so on to $y^{m-1}z$ at the end of the second row. Let the $2m+1$ special polynomials thus defined be denoted by $U_m, \cdots, U_0, V_{m-1}, \cdots, V_0$.

Then if U is any homogeneous harmonic polynomial

of the mth degree, and $c_{m00}, \cdots, c_{0,\ m-1,\ 1}$ are the first two rows of its coefficients, it is expressible in terms of $U_m, \cdots, U_0, V_{m-1}, \cdots, V_0$ in the form

$$U = \sum_{i=0}^{m} c_{m-i,i,0} U_{m-i} + \sum_{i=1}^{m} c_{m-i,i-1,1} V_{m-i}.$$

For this expression is in fact a homogeneous polynomial of the mth degree, satisfies the differential equation, and has the designated c's as coefficients of the terms formed from the first two rows of (18), and it has been shown that there can be only one polynomial meeting these specifications.

In particular, each of the $2m+1$ spherical harmonics $\rho^m u_{mn}$ and $\rho^m v_{mn}$ as a polynomial in (x, y, z) can be represented linearly in terms of $U_m, \cdots, U_0, V_{m-1}, \cdots, V_0$. If the determinant of the $(2m+1)^2$ coefficients in the representation were zero, there would be a relation of linear dependence connecting the functions $\rho^m u_{mn}, \rho^m v_{mn}$, which is impossible by the last paragraph of §2. So the equations can be solved for the U's and V's in terms of $\rho^m u_{mn}, \rho^m v_{mn}$, and any linear combination of the former set of polynomials is at the same time a linear combination of the latter. *Any homogeneous harmonic polynomial of the mth degree in (x, y, z) is a linear combination of the spherical harmonics $\rho^m u_{mn}, \rho^m v_{mn}$.*

5. Rotation of axes. A rotation of axes connecting one set of rectangular coordinates (x, y, z) with another set (ξ, η, ζ) having the same origin is described by a set of equations of the form

$$\begin{aligned}
\xi &= c_{11}x + c_{12}y + c_{13}z \\
(20) \qquad \eta &= c_{21}x + c_{22}y + c_{23}z \\
\zeta &= c_{31}x + c_{32}y + c_{33}z
\end{aligned}$$

in which

$$(21) \quad \sum_{k=1}^{3} c_{ik}^{2} = 1, \quad \sum_{k=1}^{3} c_{ik}c_{jk} = 0, \quad i, j = 1, 2, 3, \quad i \neq j,$$

while similar relations hold if the summation is performed by columns instead of rows. It will be shown that *Laplace's equation is invariant under such a transformation.*

The formulas can be written more compactly if x, y, z are replaced by x_1, x_2, x_3, and ξ, η, ζ by ξ_1, ξ_2, ξ_3. Then the equations of transformation can be summarized in the single formula

$$\xi_i = \sum_{k=1}^{3} c_{ik}x_k.$$

In the same notation.

$$\frac{\partial u}{\partial x_k} = \sum_{i=1}^{3} \frac{\partial u}{\partial \xi_i} \frac{\partial \xi_i}{\partial x_k} = \sum_{i=1}^{3} c_{ik} \frac{\partial u}{\partial \xi_i}.$$

The meaning of the expressions is naturally unchanged if some other letter, for example j, is used instead of i to represent the index of summation. By repetition of the process, with substitution of $\partial u/\partial x_k$ for u,

$$\frac{\partial^2 u}{\partial x_k^2} = \sum_{i=1}^{3} c_{ik} \frac{\partial}{\partial \xi_i}\left(\frac{\partial u}{\partial x_k}\right) = \sum_{i=1}^{3} c_{ik} \frac{\partial}{\partial \xi_i}\left(\sum_{j=1}^{3} c_{jk} \frac{\partial u}{\partial \xi_j}\right)$$

$$= \sum_{i=1}^{3} \sum_{j=1}^{3} c_{ik}c_{jk} \frac{\partial^2 u}{\partial \xi_i \partial \xi_j}.$$

Hence

$$\sum_{k=1}^{3} \frac{\partial^2 u}{\partial x_k^2} = \sum_{k=1}^{3} \sum_{i=1}^{3} \sum_{j=1}^{3} c_{ik}c_{jk} \frac{\partial^2 u}{\partial \xi_i \partial \xi_j}$$

$$= \sum_{i=1}^{3} \sum_{j=1}^{3} \left(\sum_{k=1}^{3} c_{ik}c_{jk}\right) \frac{\partial^2 u}{\partial \xi_i \partial \xi_j}.$$

Since by (21) the sum in parentheses is 1 or 0 according as i and j are the same or different, the whole triple sum reduces to $\sum \partial^2 u / \partial \xi_i^2$, or in terms of the original symbols

$$\frac{\partial^2 u}{\partial x^2} + \frac{\partial^2 u}{\partial y^2} + \frac{\partial^2 u}{\partial z^2} = \frac{\partial^2 u}{\partial \xi^2} + \frac{\partial^2 u}{\partial \eta^2} + \frac{\partial^2 u}{\partial \zeta^2}.$$

The transformation inverse to (20) is

$$x = c_{11}\xi + c_{21}\eta + c_{31}\zeta$$
$$y = c_{12}\xi + c_{22}\eta + c_{32}\zeta$$
$$z = c_{13}\xi + c_{23}\eta + c_{33}\zeta$$

with the same c's arranged by rows instead of columns. From the fact that the transformation carries a homogeneous polynomial in either set of variables into a homogeneous polynomial of the same degree in the other set, together with the fact that the form of Laplace's equation is unchanged, it follows that *a homogeneous harmonic polynomial of the mth degree in (x, y, z) is a homogeneous harmonic polynomial of the mth degree in (ξ, η, ζ), and vice versa.*

The quantity

$$\rho = (x^2 + y^2 + z^2)^{1/2} = (\xi^2 + \eta^2 + \zeta^2)^{1/2}$$

is the same in both coordinate systems. Let (ρ, γ, ψ) denote spherical coordinates corresponding to (ξ, η, ζ):

$$\xi = \rho \sin \gamma \cos \psi, \qquad \eta = \rho \sin \gamma \sin \psi, \qquad \zeta = \rho \cos \gamma.$$

Let u_{mn}, v_{mn} be used now as symbols of functional operation, with the variables indicated, i.e. let the previous u_{mn}, v_{mn} be represented more explicitly by $u_{mn}(\theta, \phi)$, $v_{mn}(\theta, \phi)$, and let $u_{mn}(\gamma, \psi)$, $v_{mn}(\gamma, \psi)$ stand for the corresponding expressions in terms of γ and ψ. Since $\rho^m u_{mn}(\gamma, \psi)$ or $\rho^m v_{mn}(\gamma, \psi)$ is a homogeneous harmonic polynomial of the mth degree in (x, y, z) as well as in (ξ, η, ζ), it can be written as a linear combination of the functions $\rho^m u_{mn}(\theta, \phi)$, $\rho^m v_{mn}(\theta, \phi)$; or by removal of the factor ρ^m, *each of the functions $u_{mn}(\gamma, \psi)$, $v_{mn}(\gamma, \psi)$ is a linear combination of the $2m+1$ spherical harmonics in (θ, ϕ) corresponding to the same value of m.* A similar statement holds with reference to the inverse transformation.

The coefficients in such a representation, which is in effect a Laplace series having only a finite number of terms different from zero, and so uncomplicated by questions of convergence, are found in the usual way by use of the properties of orthogonality of the terms, and can be written down immediately by adaptation of (17).

6. Integral representation for group of terms in the Laplace series. Let S_m denote the group of terms corresponding to a single value of m in (13):

$$(22) \quad \begin{aligned} S_m &= \tfrac{1}{2} a_{m0} u_{m0}(\theta, \phi) \\ &\quad + \sum_{n=1}^{m} \left[a_{mn} u_{mn}(\theta, \phi) + b_{mn} v_{mn}(\theta, \phi) \right]. \end{aligned}$$

(The function $u_{m0}(\theta, \phi)$ does not of course actually depend on ϕ, being $P_m(\cos\theta)$.) By (17), with (θ', ϕ') written in place of (θ, ϕ) for the variables of integration,

$$a_{mn} = \frac{(m-n)!}{(m+n)!} \frac{2m+1}{2\pi} \int_0^\pi \int_{-\pi}^\pi f(\theta', \phi') u_{mn}(\theta', \phi') \sin\theta' d\phi' d\theta',$$

$$b_{mn} = \frac{(m-n)!}{(m+n)!} \frac{2m+1}{2\pi} \int_0^\pi \int_{-\pi}^\pi f(\theta', \phi') v_{mn}(\theta', \phi') \sin\theta' d\phi' d\theta'.$$

By explicit substitution of these values in (22) it is possible to write

$$S_m = \frac{2m+1}{4\pi} \int_0^\pi \int_{-\pi}^\pi f(\theta', \phi') G_m(\theta, \phi, \theta', \phi') \sin \theta' d\phi' d\theta'$$

with

$$
\begin{aligned}
(23) \quad G_m(\theta, \phi, \theta', \phi') &= u_{m0}(\theta, \phi) u_{m0}(\theta', \phi') \\
&\quad + 2 \sum_{n=1}^m \frac{(m-n)!}{(m+n)!} \big[u_{mn}(\theta, \phi) u_{mn}(\theta', \phi') \\
&\quad\quad + v_{mn}(\theta, \phi) v_{mn}(\theta', \phi') \big].
\end{aligned}
$$

An expression more compact than this is to be found for $G_m(\theta, \phi, \theta', \phi')$.

Let points on the surface of the unit sphere be represented by their angular coordinates (colatitude and longitude), the constant $\rho = 1$ being omitted. Let (θ, ϕ) for the time being be thought of as a *fixed* point M, and let the coordinates of a variable point P be denoted by (θ', ϕ'). Let N be the "north pole," the point for which $\theta = 0$. In the spherical triangle MNP, the sides NM and NP are θ and θ' respectively, and the angle N (apart from algebraic sign, which is immaterial for present purposes) is $\phi' - \phi$. Let the angular measure of the side MP be denoted by γ. Then by the law of cosines in spherical trigonometry

$$(24) \quad \cos \gamma = \cos \theta \cos \theta' + \sin \theta \sin \theta' \cos (\phi' - \phi).$$

Let a new system of rectangular coordinates (ξ, η, ζ) be set up with the ζ-axis passing through M. Let (γ, ψ) be the corresponding coordinates of P on the surface of the sphere, this definition of γ being in agreement with the one in the preceding paragraph. By §5, applied to the coordinates (θ', ϕ') and (γ, ψ), the function $u_{m0}(\gamma, \psi) = P_m(\cos \gamma)$ can be written in the form

$$(25) \quad \tfrac{1}{2}A_0 u_{m0}(\theta', \phi') + \sum_{n=1}^{m} \left[A_n u_{mn}(\theta', \phi') + B_n v_{mn}(\theta', \phi') \right]$$

with

$$
(26) \quad
\begin{aligned}
A_n &= \frac{(m - n)!}{(m + n)!} \frac{2m + 1}{2\pi} \times \\
&\int_0^\pi \int_{-\pi}^\pi P_m(\cos\gamma) u_{mn}(\theta', \phi') \sin\theta' d\phi' d\theta'
\end{aligned}
$$

and a corresponding formula for B_n. On the other hand, by a converse application of the same principle, each $u_{mn}(\theta', \phi')$ and $v_{mn}(\theta', \phi')$ can be represented by an expression

$$(27) \quad \tfrac{1}{2}\alpha_0 u_{m0}(\gamma, \psi) + \sum_{k=1}^{m} \left[\alpha_k u_{mk}(\gamma, \psi) + \beta_k v_{mk}(\gamma, \psi) \right],$$

with coefficients varying, of course, from one function to another of the set in question.

If $F(\theta', \phi') = H(\gamma, \psi)$ is any function expressed alternatively in terms of the two coordinate systems,

$$\int_0^\pi \int_{-\pi}^\pi F(\theta', \phi') \sin\theta' d\phi' d\theta' = \int_0^\pi \int_{-\pi}^\pi H(\gamma, \psi) \sin\gamma \, d\psi \, d\gamma,$$

either formula representing integration of the function over the whole surface of the sphere. Let (27), assumed for the moment to represent $u_{mn}(\theta', \phi')$, be substituted under the sign of integration in (26), and let the integration be thought of as performed with respect to γ and ψ. Since $P_m(\cos\gamma)$ is orthogonal to each of the functions $u_{mk}(\gamma, \psi)$, $v_{mk}(\gamma, \psi)$ for $k \geq 1$, and since $u_{m0}(\gamma, \psi) = P_m(\cos\gamma)$,

$$\int_0^\pi \int_{-\pi}^\pi P_m(\cos\gamma) u_{mn}(\theta', \phi') \sin\theta' d\phi' d\theta'$$

$$= \int_0^\pi \int_{-\pi}^\pi \tfrac{1}{2}\alpha_0 [P_m(\cos\gamma)]^2 \sin\gamma \, d\psi \, d\gamma$$

$$= \tfrac{1}{2}\alpha_0 \int_{-\pi}^\pi d\psi \int_{-1}^1 [P_m(x)]^2 dx = \frac{2\pi\alpha_0}{2m+1},$$

so that

$$A_n = \frac{(m-n)!}{(m+n)!} \alpha_0.$$

But the value of α_0 can be found explicitly by setting $\gamma = 0$ in (27), which is still understood to be a representation of $u_{mn}(\theta', \phi')$. When $\gamma = 0$, θ' and ϕ' reduce to θ and ϕ; $u_{m0}(\gamma, \psi) = P_m(\cos\gamma)$ has the value 1; and each term under the sign of summation vanishes, having a positive power of $\sin\gamma$ as a factor. So

$$\tfrac{1}{2}\alpha_0 = u_{mn}(\theta, \phi), \qquad A_n = 2\frac{(m-n)!}{(m+n)!} u_{mn}(\theta, \phi).$$

Similarly, by substitution of $v_{mn}(\theta', \phi')$ for $u_{mn}(\theta', \phi')$,

$$B_n = 2\frac{(m-n)!}{(m+n)!} v_{mn}(\theta, \phi).$$

The calculation of A_n is valid in particular for $n = 0$: $A_0 = 2u_{m0}(\theta, \phi)$.

On substitution of these coefficients in the representation (25) of $P_m(\cos\gamma)$ and comparison with the expression for G_m in (23) it is seen that the two are identical. Consequently

$$S_m = \frac{2m+1}{4\pi} \int_0^\pi \int_{-\pi}^\pi f(\theta', \phi') P_m(\cos\gamma) \sin\theta' d\phi' d\theta',$$

$\cos \gamma$ being given by (24). When S_m has been calculated, the Laplace series (13) is merely $\sum_0^\infty S_m$.

As an application of this result, the formula corresponding in three dimensions to Poisson's integral in §6 of Chapter IV can readily be derived. The function $u(\rho, \theta, \phi)$ in the formula preceding (13), the solution of the boundary value problem in the sphere, is $\sum_0^\infty \rho^m S_m$, which is now seen, formally at least, to have the representation

$$u(\rho, \theta, \phi) = \frac{1}{4\pi} \int_0^\pi \int_{-\pi}^\pi f(\theta', \phi') \Phi(\rho, \theta, \phi, \theta', \phi') \sin \theta' d\phi' d\theta',$$

where

$$\Phi(\rho, \theta, \phi, \theta', \phi') = \sum_{m=0}^\infty (2m + 1)\rho^m P_m(\cos \gamma).$$

By (1) and (2) of Chapter II,

$$\sum_{m=0}^\infty \rho^m P_m(\cos \gamma) = H(\cos \gamma, \rho) = (1 - 2\rho \cos \gamma + \rho^2)^{-1/2}.$$

the series being convergent for all values of γ when $0 \leqq \rho < 1$, since $\left| P_m(x) \right| \leqq 1$ for $\left| x \right| \leqq 1$; and

$$\sum_{m=0}^\infty m\rho^m P_m(\cos \gamma) = \rho(\partial/\partial\rho)H(\cos \gamma, \rho)$$
$$= (\rho \cos \gamma - \rho^2)(1 - 2\rho \cos \gamma + \rho^2)^{-3/2}.$$

Thus the representations of Φ and u take the form

$$\Phi(\rho, \theta, \phi, \theta', \phi') = (1 - \rho^2)(1 - 2\rho \cos \gamma + \rho^2)^{-3/2},$$
$$u(\rho, \theta, \phi)$$
$$= \frac{1}{4\pi} \int_0^\pi \int_{-\pi}^\pi f(\theta', \phi') \frac{1 - \rho^2}{(1 - 2\rho \cos \gamma + \rho^2)^{3/2}} \sin \theta' d\phi' d\theta'.$$

This is Poisson's integral in three dimensions. It may be noted that if M' is the point inside the sphere with spherical coordinates (ρ, θ, ϕ), and P the point on the surface with coordinates $(1, \theta', \phi')$, $1 - 2\rho \cos \gamma + \rho^2$ is the square of the distance $M'P$. A similar interpretation is possible in (24) of Chapter IV.

7. Completeness of the Laplace series. It may naturally be inquired what reason there is for anticipating that the set of particular functions from which the terms of the Laplace series are formed will be sufficiently general in the aggregate to serve for the representation of an arbitrary function on the surface of the sphere. A complete discussion of this question would not be in place here, and would in particular require a careful statement of limitations on the type of "arbitrariness" to be admitted, but conclusions of a high degree of generality will be apparent if the following observations are accepted as self-evident or plausible:

a) If $f(\theta, \phi)$ is a continuous function on the surface of the unit sphere, $\rho f(\theta, \phi)$ is a continuous function in three dimensions;

b) By Weierstrass's theorem on polynomial approximation as formulated for functions of three variables, any continuous function in a bounded region of space can be uniformly approximated with any assigned degree of accuracy by polynomials in (x, y, z); such approximation throughout a three-dimensional region including the unit sphere then yields in particular an approximation on the surface of the sphere;

c) The values of any polynomial on the surface of the sphere can be expressed by means of the identity $x^2 + y^2 + z^2 = 1$ as surface values of a polynomial containing no power of z higher than the first;

d) In particular, each of the

$$1 + 3 + 5 + \cdots + (2m + 1) = (m + 1)^2$$

homogeneous harmonic polynomials of degree $\leq m$ can be expressed on the surface as a linear combination of the $(m+1)^2$ monomials $x^i y^j z^k$ in which $i+j+k \leq m$ and $k \leq 1$;

e) If the determinant of the $(m+1)^4$ coefficients in this representation were zero, there would be a linear relation connecting the values of the polynomials on the surface, i.e. connecting the spherical harmonics in θ and ϕ to which they reduce there;

f) There can be no linear relation connecting any finite number of the functions u_{mn}, v_{mn}, whether with the same or different values of m; for if one of them were linearly expressible in terms of the others, being orthogonal to each of the others, it would be orthogonal to itself, and so identically zero;

g) The $(m+1)^2$ equations in d) can be solved for the monomials in terms of the homogeneous harmonic polynomials;

h) Any polynomial, and in particular the approximating polynomial in b), can be linearly expressed on the surface of the sphere in terms of homogeneous harmonic polynomials, i.e. in terms of the spherical harmonics u_{mn}, v_{mn} with various values of m.

That is to say, in summary, any continuous function on the surface of the sphere can be uniformly approximated there by means of spherical harmonics with any desired degree of accuracy.

8. Boundary value problem in a cylinder; series involving Bessel functions of positive order.

In the first boundary value problem of §9 in Chapter IV, if the functions are not supposed to be independent of ϕ, (30) and (31) are to be replaced by

$$(28) \qquad \frac{\partial^2 u}{\partial r^2} + \frac{1}{r} \frac{\partial u}{\partial r} + \frac{1}{r^2} \frac{\partial^2 u}{\partial \phi^2} + \frac{\partial^2 u}{\partial z^2} = 0,$$

$$(29) \quad u(1, \phi, z) = 0, \quad u(r, \phi, b) = 0, \quad u(r, \phi, 0) = f(r, \phi).$$

The usual procedure, with substitution of a product $R(r) \, F(\phi) \, Z(z)$ for u in (28) and with notation for the constants as indicated, leads to the formulas

$$R''FZ + \frac{1}{r} R'FZ + \frac{1}{r^2} RF''Z + RFZ'' = 0,$$

$$\frac{R'' + (1/r)R'}{R} + \frac{1}{r^2} \frac{F''}{F} = -\frac{Z''}{Z} = -\lambda^2,$$

$$\frac{r^2R'' + rR'}{R} + \lambda^2 r^2 = -\frac{F''}{F} = n^2,$$

$$(30) \qquad R'' + \frac{1}{r} R' + \left(\lambda^2 - \frac{n^2}{r^2}\right) R = 0.$$

The equation $Z'' = \lambda^2 Z$ has as a combination of $e^{\lambda z}$ and $e^{-\lambda z}$ the solution $\sinh \lambda(b - z)$, which vanishes for $z = b$. The equation $F'' = -n^2 F$ has the solutions $\cos n\phi$ and $\sin n\phi$, which for integral n remain unchanged on replacement of one value of the ϕ-coordinate by another belonging to the same point and so differing from the first by an integral multiple of 2π. By §8 of Chapter III the equation (30) has for a solution $J_n(\lambda r)$.

If $\lambda_{n1}, \lambda_{n2}, \cdots$ are the positive roots of the equation $J_n(\lambda) = 0$, (28) and the first two conditions of (29) are satisfied by

$$\sinh \lambda_{nk}(b - z) \cos n\phi J_n(\lambda_{nk} r), \; \sinh \lambda_{nk}(b - z) \sin n\phi J_n(\lambda_{nk} r)$$

for $n = 0, 1, 2, \cdots$, $k = 1, 2, \cdots$, the value $n = 0$ being omitted, of course, in the case of the sine. The discussion continues with the formulas

$$u(r, \phi, z) = \frac{1}{2} \sum_{k=1}^{\infty} \alpha_{0k} \sinh \lambda_{0k}(b - z)J_0(\lambda_{0k}r)$$

$$+ \sum_{n=1}^{\infty} \sum_{k=1}^{\infty} \sinh \lambda_{nk}(b - z)(\alpha_{nk} \cos n\phi$$

$$+ \beta_{nk} \sin n\phi)J_n(\lambda_{nk}r),$$

$$u(r, \phi, 0) = f(r, \phi) = \frac{1}{2} \sum_{k=1}^{\infty} a_{0k}J_0(\lambda_{0k}r)$$

(31)

$$+ \sum_{n=1}^{\infty} \sum_{k=1}^{\infty} (a_{nk} \cos n\phi + b_{nk} \sin n\phi)J_n(\lambda_{nk}r),$$

$$\alpha_{nk} = a_{nk}/\sinh \lambda_{nk}b, \qquad \beta_{nk} = b_{nk}/\sinh \lambda_{nk}b.$$

The functions $\cos n\phi \ J_n(\lambda_{nk}r)$ and $\sin n\phi \ J_n(\lambda_{nk}r)$, considered for all the pairs of subscripts which enter into (31), form an orthogonal set, in the sense that the integral of the product of any two distinct functions of the set over the unit circle in the (r, ϕ)-plane is zero. If $F(r, \phi)$ is any function of these variables its integral over the circle is

$$\int_{-\pi}^{\pi} \int_0^1 F(r, \phi) \ r \, dr \, d\phi.$$

For $F(r, \phi) = \cos n\phi J_n(\lambda_{nk}r) \cos m\phi J_m(\lambda_{mh}r)$ the integral reduces to the product

$$(32) \qquad \int_{-\pi}^{\pi} \cos n\phi \cos m\phi \, d\phi \int_0^1 r J_n(\lambda_{nk}r)J_m(\lambda_{mh}r)dr.$$

Here the first factor is zero if $m \neq n$, and the second factor is zero, by §8 of Chapter III, if $m = n$, $h \neq k$. There is a corresponding calculation, of course, for the products involving $\sin n\phi \sin m\phi$ or $\cos n\phi \sin m\phi$.

If d_{nk} denotes the value of the second integral in (32) when $m = n$, $h = k$ (see (19) in Chapter III), the coefficients in (31) are

$$a_{nk} = \frac{1}{\pi d_{nk}} \int_{-\pi}^{\pi} \int_{0}^{1} f(r, \phi) \cos n\phi J_n(\lambda_{nk}r) \, r \, dr \, d\phi,$$

$$b_{nk} = \frac{1}{\pi d_{nk}} \int_{-\pi}^{\pi} \int_{0}^{1} f(r, \phi) \sin n\phi J_n(\lambda_{nk}r) \, r \, dr \, d\phi,$$

the formula for a_{nk} being valid in particular for $n = 0$.

It may be noted that the terms of (31) having $\cos n\phi$ as a common factor, for a fixed value of $n > 0$, are obtained by multiplying by $\cos n\phi$ the expansion of the function

$$f_n(r) = \frac{1}{\pi} \int_{-\pi}^{\pi} f(r, \phi) \cos n\phi \, d\phi$$

in a simple series of the functions $J_n(\lambda_{nk}r)$, with a similar remark for the terms involving $\sin n\phi$. The double series may therefore be regarded as involving incidentally an application of this type of expansion as discussed in Chapter III.

SUPPLEMENTARY REFERENCES: Byerly; Churchill; Whittaker and Watson, Chapter XVIII; Hobson (2); Gray, Mathews, and MacRobert; Kellogg; Courant-Hilbert; Riemann-Weber.

CHAPTER VI

THE PEARSON FREQUENCY FUNCTIONS

1. The Pearson differential equation. The subject matter of this chapter has at first sight no obvious connection with what has gone before, but is introduced here to serve as a background for subsequent developments (see §10 in Chapter VII).

A class of functions proposed by Karl Pearson for the representation of statistical frequencies, and extensively used for that purpose, is defined by the differential equation

$$(1) \qquad \frac{1}{y} \frac{dy}{dx} = \frac{D + Ex}{A + Bx + Cx^2},$$

in which A, B, C, D, E are constants. The "normal" frequency function $e^{-x^2/2}$ satisfies this equation with $E = -1$, $A = 1$, $B = C = D = 0$. The Pearson functions may be regarded as a generalization starting from this as a particular case. The resulting types of function will obviously be subject to classification depending on the characteristics of the denominator $A + Bx + Cx^2$, namely whether it is actually of the second or of lower degree, and, as regards behavior for real values of the variable, whether the roots of the equation obtained by setting it equal to zero are real or complex.

2. Quadratic denominator, real roots. Let it be supposed first that $C \neq 0$, and that the equation

$$(2) \qquad A + Bx + Cx^2 = 0$$

has distinct real roots a, b, with $a > b$. After division of numerator and denominator in the right-hand member of (1) by the non-vanishing constant C, the differential equation can be written in terms of partial fractions in the form

$$(3) \qquad \frac{1}{y} \frac{dy}{dx} = \frac{\alpha}{x - a} + \frac{\beta}{x - b} = \frac{\beta}{x - b} - \frac{\alpha}{a - x},$$

α and β being constants.

The expression in the third member of (3) being preferred for initial consideration because both denominators are positive in the interval $b < x < a$, the general solution is given by

$$\log y = \beta \log (x - b) + \alpha \log (a - x) + K,$$
$$y = k(a - x)^{\alpha}(x - b)^{\beta}.$$

This function is real in the interval (b, a) (when k is real) for arbitrary real exponents α, β. It takes on the value 0 for $x = a$ if $\alpha > 0$, takes on a value different from zero if $\alpha = 0$, and becomes infinite if $\alpha < 0$. (It is assumed that $k \neq 0$.) There are corresponding types of behavior at the other end of the interval. For the interpretation as a frequency function, and also for the application to the theory of orthogonal polynomials which is important for the present exposition, y must be integrable over the interval (b, a). This requires that $\alpha > -1$, $\beta > -1$, a divergent improper integral being obtained otherwise. In the applications furthermore the factor k will be positive.

The derivative dy/dx vanishes when x has the value

$$x_0 = (\alpha b + \beta a)/(\alpha + \beta).$$

If α and β are both positive or both negative, $b < x_0 < a$. In the former case y vanishes at a and b and is positive between (for $k > 0$) and consequently has a maximum in the interior of the interval, which must be at the point $x = x_0$. If the exponents are negative, $x = x_0$ gives a minimum. Let $x' = x - x_0$, $c = a - x_0$, $d = x_0 - b$; then the function becomes

$$y = y_0 \left(1 - \frac{x'}{c}\right)^\alpha \left(1 + \frac{x'}{d}\right)^\beta,$$

with $y_0 = kc^\alpha d^\beta$, and $c/d = \alpha/\beta$; here y_0 is the maximum (or minimum) ordinate in the interval, attained for $x' = 0$.

Without regard to the maximum or minimum, and without regard to the values of α and β (which may or may not now be of the same sign), the interval (b, a) can be reduced by a linear transformation of x, i.e. a transformation of the form $x' = gx + h$, to the interval $(-1, 1)$. That is to say, under the present hypothesis of distinct real roots, the denominator in (1) can without essential loss of generality be taken at the outset as $1 - x^2$. This form of specialization will be preferred in the subsequent discussion of orthogonal polynomials.

Integration of (3) on the basis of the form of representation in the second member instead of the third gives

$$y = k(x - a)^\alpha (x - b)^\beta.$$

If $b < a$ this function is real for arbitrary α and β on the infinite interval $x > a$. For integrability at the lower end of this interval it is requisite that $\alpha > -1$. As x becomes infinite, y is of the order of magnitude of $x^{\alpha+\beta}$, and the requirement for convergence when the integration is extended to infinity is that $\alpha + \beta < -1$. For the statistical applications it is necessary also that $\int_a^\infty x^p y \, dx$ exist for a certain number of positive integral values of

p, i.e. that the frequency distribution represented by the function y have finite "moments" of corresponding orders. For finiteness of the pth moment it is necessary that $\alpha + \beta + p < -1$.

A third form of solution coordinate with the two preceding, but not different from the second in its essential characteristics, is $y = k(a-x)^{\alpha}(b-x)^{\beta}$, real for arbitrary α and β on the interval $(-\infty, b)$ if $b < a$.

In the case that (2) has equal roots it may be supposed without essential loss of generality that the double root occurs for $x = 0$, since this can be brought about by a linear transformation of x. Then (1) has the form

$$\frac{1}{y} \frac{dy}{dx} = \frac{\alpha}{x^2} + \frac{\beta}{x}.$$

Integration gives

$$\log y = -\alpha/x + \beta \log x + K,$$
$$y = k x^{\beta} e^{-\alpha/x}.$$

This is real for arbitrary β (and arbitrary α) when $x > 0$. The exponential factor approaches 1 as x becomes infinite, and convergence of the integral of the function over an interval extending to infinity demands that $\beta < -1$; for a finite pth moment, $\beta + p < -1$. Then y is not integrable over an interval extending to the right from $x = 0$ if $\alpha \leqq 0$, but is integrable if $\alpha > 0$, without further restriction on β.

3. Quadratic denominator, complex roots. When the roots of (2) are complex the differential equation, after a preliminary change of origin for x if necessary, can be taken as

$$\frac{1}{y} \frac{dy}{dx} = -\frac{a\alpha + 2\beta x}{x^2 + a^2};$$

the value of $a > 0$ in the denominator being given, the $-$ sign and the coefficients a and 2 in the numerator are arbitrary items of notation which affect merely the definition of the constants α and β. The solution then is

$$\log y = -\,\alpha \text{ arc tan } (x/a) - \beta \log (x^2 + a^2) + K,$$

$$y = \frac{k e^{-\alpha \text{ arc tan } (x/a)}}{(x^2 + a^2)^\beta}.$$

This is real for all real values of x if the constants are real. The exponential function in the numerator is continuous for all values of x, and approaches the limits $e^{\pm \alpha \pi / 2}$ as x becomes negatively or positively infinite, and so has no effect one way or the other on convergence of the integral of y from $-\infty$ to ∞. The form of y is of course particularly simple if $\alpha = 0$. As x becomes infinite in either direction, y is of the order of magnitude of $x^{-2\beta}$. The requirement for convergence of the integral over the infinite interval is that $\beta > \frac{1}{2}$, and for finiteness of the pth moment, that $\beta > (p+1)/2$.

4. Linear or constant denominator. If $C = 0$, $B \neq 0$ in (1), it is possible by a change of origin for x to replace A by 0. The differential equation then takes the form

$$\frac{1}{y} \frac{dy}{dx} = \frac{\alpha}{x} - \beta.$$

The solution is

$$\log y = \alpha \log x - \beta x + K,$$
$$y = k x^\alpha e^{-\beta x}.$$

This is real for $x > 0$, the constants being supposed real. The condition for integrability near $x = 0$ is that $\alpha > -1$, and integrability to infinity then requires that $\beta > 0$.

When these conditions are satisfied the integrals defining moments of all positive orders are convergent.

By a further linear transformation of x, leaving the origin unaltered but changing the scale of measurement along the x-axis, the coefficient β in the exponent can be replaced by 1. The function then is except for a constant factor the integrand in the ordinary integral representation of $\Gamma(\alpha+1)$. On the other hand, a superficially more general but essentially equivalent form would have been obtained if the constant term had not been removed from the denominator in the differential equation by shifting the origin at the outset.

If $B = C = 0$, simplification of the numerator in (1) by a change of origin reduces the differential equation to

$$\frac{1}{y}\frac{dy}{dx} = -\beta x,$$

with solution $\log y = -\tfrac{1}{2}\beta x^2 + K$, $y = ke^{-\beta x^2/2}$. With real constants this function is real for all real values of x. For integrability over the interval $(-\infty, \infty)$ it is necessary that $\beta > 0$; when $\beta > 0$ the moments of all orders are finite. By a change of scale for x the value of β can be reduced to 1, and y is then a constant multiple of the normal frequency function in the particular form referred to in §1.

5. Finiteness of moments. It will be seen in later chapters that functions satisfying the Pearson differential equation (1) are of particular importance as "weight functions" for systems of orthogonal polynomials on the appropriate intervals. To serve in this connection, however, a function must not only be integrable itself over the interval in question, but must be such that when it is multiplied by an arbitrary polynomial the product is integrable over the interval. That is to say, the corresponding moments of all orders must be finite. In the case of a finite interval this requirement imposes no additional restriction; but it is fulfilled only by the functions of §4 among those for which the interval is infinite. There are then just three types of solution to be considered further in the present discussion: that of §2 with a finite interval, that of §4 with an interval infinite in one direction, and the other type in §4 with an interval infinite in both directions. To these types correspond respectively the Jacobi, Laguerre, and Hermite polynomials. It is to be noted that in each of the cases specified the product of the function by the denominator $A + Bx + Cx^2$ vanishes at the ends of the interval, whether this is finite or infinite in extent.

SUPPLEMENTARY REFERENCES: Elderton; Pearson; Rietz.

CHAPTER VII

ORTHOGONAL POLYNOMIALS

1. Weight function. It is seen by reference to §3 of Chapter V that

$$\int_{-1}^{1} (1 - x^2) P_m'(x) P_k'(x) dx = 0$$

if $P_m(x)$, $P_k(x)$ are Legendre polynomials of different degrees. This relation can be described by saying that the functions $(1-x^2)^{1/2} P_m'(x)$, $(1-x^2)^{1/2} P_k'(x)$ are orthogonal to each other over the interval $(-1, 1)$. It can be alternatively expressed, in a more significant manner for the following discussion, by saying that the polynomials $P_m'(x)$, $P_k'(x)$ themselves are orthogonal *with respect to the weight function* $1-x^2$. More generally, the polynomials $P_m^{(n)}(x)$, $P_k^{(n)}(x)$ are orthogonal with respect to $(1-x^2)^n$ as weight function, if n is any positive integer. The concept of systems of polynomials orthogonal with respect to a weight function will be dominant from now on.

In the relation

$$(1) \qquad \int_{0}^{\pi} \cos m\theta \cos k\theta \, d\theta = 0, \qquad m \neq k,$$

let $x = \cos \theta$, $dx = -(1-x^2)^{1/2} d\theta$. The function $\cos m\theta$ can be expressed as a polynomial of the mth degree in $\cos \theta$:

$$\cos m\theta = C_m(\cos \theta) = C_m(x),$$

and similarly for cos $k\theta$. Thus (1) takes the form

$$\int_{-1}^{1} (1 - x^2)^{-1/2} C_m(x) C_k(x) dx = 0.$$

The polynomials $C_m(x)$, $m = 0, 1, 2, \cdots$, are orthogonal for weight function $(1 - x^2)^{-1/2}$. (The designation of the interval of integration, which is of course essential for completeness of statement, will not be repeated in every instance when it has once been made clear.) The function $\sin (m+1)\theta$ has the expression

$$\sin (m + 1)\theta = \sin \theta \, S_m(\cos \theta),$$

where S_m is a polynomial of the mth degree. The relation

$$\int_{0}^{\pi} \sin (m + 1)\theta \sin (k + 1)\theta \, d\theta = 0, \quad m \neq k,$$

is equivalent to

$$\int_{-1}^{1} (1 - x^2)^{1/2} S_m(x) S_k(x) dx = 0.$$

The polynomials $S_m(x)$ are orthogonal for weight $(1 - x^2)^{1/2}$.

The polynomials $C_m(x)$ and $S_m(x)$ are called trigonometric polynomials or Tchebichef polynomials. Together with the Legendre polynomials and the derivatives of the Legendre polynomials they are particular cases of Jacobi polynomials, one of three important types to which special attention will be given in later chapters. The name *Tchebichef polynomials* is also applied more generally to systems of polynomials orthogonal with repect to an arbitrary weight function.* The present

---

* In the theory of the approximate representation of a given function by means of polynomials, the designation *Tchebichef polynomial*

chapter is concerned with the beginnings of a general theory of such orthogonal systems.

2. Schmidt's process. Let $\phi_0(x)$, $\phi_1(x)$, $\phi_2(x)$, $\cdots$ be an arbitrary sequence of functions on an interval (a, b), subject merely to conditions of integrability and linear independence. Throughout this chapter the interval (a, b) may be replaced by an infinite interval (a, ∞) or by the interval $(-\infty, \infty)$, provided that all the integrals in question exist. It is assumed that every linear combination $\psi = k_0\phi_0 + \cdots + k_m\phi_m$ of a finite number of the ϕ's, with constant coefficients $k_0, \cdots, k_m$ not all zero, is different from zero on a set of points sufficient to make the definite integral of ψ^2 over the interval different from zero (and so necessarily positive).

Let

$$d_0 = \int_a^b \left[\phi_0(x)\right]^2 dx, \qquad g_0(x) = \phi_0(x)/d_0^{1/2},$$

so that $\int_a^b [g_0(x)]^2 dx = 1$. Let

$$G_1(x) = \phi_1(x) - c_{10}g_0(x), \qquad c_{10} = \int_a^b \phi_1(x)g_0(x)dx,$$

$$d_1 = \int_a^b \left[G_1(x)\right]^2 dx, \qquad g_1(x) = G_1(x)/d_1^{1/2}.$$

Then

$$\int_a^b G_1 g_0 dx = 0 = \int_a^b g_1 g_0 dx, \qquad \int_a^b g_1^2 dx = 1.$$

In general, let functions $g_2(x)$, $g_3(x)$, $\cdots$ be defined successively by the relations

---

is used in a still different sense to refer to the approximating polynomial of specified degree for which the maximum error is as small as possible.

$$G_n(x) = \phi_n(x) - \sum_{k=0}^{n-1} c_{nk} g_k(x), \qquad c_{nk} = \int_a^b \phi_n(x) g_k(x) dx$$

$$d_n = \int_a^b [G_n(x)]^2 dx, \qquad g_n(x) = G_n(x)/d_n^{1/2}.$$

It follows immediately from the definition that each g_n is *orthogonal* to $g_0, \cdots, g_{n-1}$, and is *normalized*, i.e. has 1 for the integral of its square over the interval (a, b). Since the relation of orthogonality is symmetric, it can be said equally well that any two of the g's are orthogonal to each other. The fact that the g's are both orthogonal and normalized is summarized in a single word by calling them *orthonormal*. If the ϕ's in particular are the functions 1, x, x^2, $\cdots$ on the interval $(-1, 1)$, the corresponding g's are the normalized Legendre polynomials (see §6 of Chapter II).

This procedure for constructing an orthogonal system from an arbitrarily given set of functions is commonly referred to as *Schmidt's process of orthogonalization.*[*]

The function g_n is a linear combination of $\phi_0, \cdots, \phi_n$ (the phrase *linear combination* will be understood always to mean *linear combination with constant coefficients*), and conversely, since the coefficient of ϕ_n in the expression for g_n is different from zero in each case, the relations connecting the ϕ's with the g's can be solved successively for the ϕ's, and ϕ_n is a linear combination of $g_0, \cdots, g_n$. It follows that g_n is orthogonal to every linear combination of $\phi_0, \cdots, \phi_{n-1}$.

[*] See E. Schmidt, *Zur Theorie der linearen und nichtlinearen Integralgleichungen.* I. Teil: *Entwicklung willkürlicher Funktionen nach Systemen vorgeschriebener*, Mathematische Annalen, vol. 63 (1907), pp. 433–476; pp. 442–444.

If $\gamma(x)$ is any linear combination of $\phi_0, \cdots, \phi_n$ which is orthogonal to each of the functions $\phi_0, \cdots, \phi_{n-1}$, it must be a constant multiple of $g_n(x)$. This is apparent from the process of construction of $g_n(x)$, in which c_{10} and the subsequent coefficients c_{nk} in the functions $G_n(x)$ are uniquely determined by the requirement of orthogonality at each step, the coefficient of $\phi_n(x)$ being unity, although the property of orthogonality taken by itself admits multiplication of the whole expression $G_n(x)$ by an arbitrary constant factor; or it can be proved at a single stroke as follows. If $a_n > 0$ and a_n' are the coefficients of ϕ_n in g_n and γ respectively, the expression $\gamma - (a_n'/a_n)g_n$ does not contain ϕ_n; it is a linear combination of $\phi_0, \cdots, \phi_{n-1}$ which is orthogonal to each of these functions, and so orthogonal to itself, i.e. the integral of its square over the interval is zero, and this signifies that the coefficients by means of which it is expressed in terms of $\phi_0, \cdots, \phi_{n-1}$ must all be zero, by the hypothesis of linear independence of the ϕ's. If it is further specified that γ is normalized and $a_n' \geqq 0$, γ must be identical with g_n.

3. Orthogonal polynomials corresponding to an arbitrary weight function. Let $\rho(x)$ be a non-negative function which is integrable over (a, b), and is actually positive on a set of points such that its definite integral over (a, b) is positive. In the most important applications $\rho(x)$ will be continuous and positive throughout the interval except possibly at the end points, where it may vanish or may become infinite. In the case of an infinite interval it is to be assumed that the product of $\rho(x)$ by an arbitrary polynomial is integrable over the interval. Let the products $[\rho(x)]^{1/2}x^k$, $k = 0, 1, 2, \cdots$ be taken as the functions $\phi_k(x)$ of the preceding section.

The corresponding functions $g_n(x)$, being linear combinations of these, will be of the form $[\rho(x)]^{1/2}p_n(x)$, where $p_n(x)$ is a polynomial.

The polynomials $p_n(x)$ are the *normalized orthogonal* or *orthonormal* polynomials with $\rho(x)$ as weight function. They satisfy the conditions

$$\int_a^b \rho(x)p_m(x)p_n(x)dx = 0, \qquad m \neq n,$$

$$\int_a^b \rho(x)[p_n(x)]^2dx = 1.$$

Each is of the degree indicated by its subscript, the term in x^n having a positive coefficient. These properties determine the system of polynomials $p_n(x)$ completely.

Every polynomial of the nth degree can be expressed as a linear combination of $p_0(x), \cdots, p_n(x)$. Each $p_n(x)$ is orthogonal to every polynomial of lower degree with respect to the weight function $\rho(x)$, i.e. if $q(x)$ is any such polynomial,

$$\int_a^b \rho(x)p_n(x)q(x)dx = 0.$$

These facts are immediate corollaries of the general results previously obtained.

Some of the properties of Legendre polynomials developed in Chapter II carry over to general systems of orthogonal polynomials. Others are associated only with particular types of weight function, while still others, of course, are peculiar to the Legendre polynomials themselves. Some of the generalizations, complete or partial, will be brought out in the following pages.

4. Development of an arbitrary function in series.
The polynomials $p_n(x)$ can be used for the formal expansion of an arbitrary function in series in essentially the same way as the orthogonal functions of the earlier chapters. The formulas are complicated by the presence of the weight function, but are simplified, on the other hand, by the fact that the p's are normalized. In the relation

$$(2) \qquad f(x) = c_0 p_0(x) + c_1 p_1(x) + \cdots$$

multiplication by $\rho(x)\ p_k(x)$, followed by integration from a to b, gives

$$(3) \qquad c_k = \int_a^b \rho(x) f(x) p_k(x) dx.$$

If $f(x)$ is any function such that the integral in (3) exists for each value of k, this formula can be used to define a sequence of coefficients c_k, and in analogy with the cases previously considered the series $\sum c_k p_k(x)$ can then be studied on its own merits, without *a priori* assumption as to its convergence.

Let $s_n(x)$ denote the partial sum of the series in (2) through terms of the nth degree:

$$(4) \qquad s_n(x) = c_0 p_0(x) + c_1 p_1(x) + \cdots + c_n p_n(x).$$

If t is used as variable of integration in place of x in (3), and if the resulting expressions for the c's are substituted explicitly in (4), it is found that

$$(5) \qquad s_n(x) = \int_a^b \rho(t) f(t) K_n(x, t) dt,$$

with

$$(6) \qquad K_n(x, t) = K_n(t, x) = \sum_{k=0}^n p_k(t) p_k(x).$$

In particular, if $f(x)$ itself is a polynomial of the nth or lower degree, $f(x) = \pi_n(x)$, it is known from the preceding section that a representation of the form (2) exists, the right-hand member being a finite sum instead of an infinite series. The procedure for determining the coefficients then applies without question of convergence, and the coefficients are given by (3). In this case $s_n(x)$ is the same as $\pi_n(x)$, and $\pi_n(x)$ is reproduced identically by the formula

$$(7) \qquad \pi_n(x) = \int_a^b \rho(t)\pi_n(t)K_n(x, t)dt.$$

5. Formula of recurrence. The product $xp_n(x)$, as a polynomial of the $(n+1)$st degree, is expressible in the form

$$(8) \qquad xp_n(x) = \sum_{k=0}^{n+1} c_{nk}p_k(x),$$

with

$$(9) \qquad c_{nk} = \int_a^b \rho(x)xp_n(x)p_k(x)dx.$$

If $k < n-1$, $xp_k(x)$ is a polynomial of degree $k+1 < n$, and since $p_n(x)$ is orthogonal to every such polynomial with respect to the weight function ρ, all the coefficients c_{nk} with $k < n-1$ vanish. *Let a_k denote the coefficient of x^k in $p_k(x)$ for each value of k.* Comparison of the coefficients of x^{n+1} in (8) gives $c_{n,n+1} = a_n/a_{n+1}$. Since furthermore $c_{nk} = c_{kn}$ for all values of n and k, by (9), $c_{n,n-1} = c_{n-1,n} = a_{n-1}/a_n$. Thus (8) reduces to the following *formula of recurrence* connecting any three successive p's:

$$(10) \quad xp_n(x) = \frac{a_n}{a_{n+1}} p_{n+1}(x) + c_{nn}p_n(x) + \frac{a_{n-1}}{a_n} p_{n-1}(x).$$

If for convenience of notation a symbol $p_{-1}(x)$ is introduced and defined as having the value 0 identically, with $a_{-1}=0$, (10) may be considered to hold for $n=0$ as well as for positive values of n.

If b_k denotes the coefficient of x^{k-1} in $p_k(x)$, so that $p_k(x)=a_k x^k+b_k x^{k-1}+ \cdots$ for each k, c_{nn} is found by comparison of the coefficients of x^n in (10) to have the value

$$c_{nn} = \frac{b_n}{a_n} - \frac{b_{n+1}}{a_{n+1}} .$$

6. Christoffel-Darboux identity. Let (10) be multiplied by $p_n(t)$:

$$xp_n(x)p_n(t) = \frac{a_n}{a_{n+1}} p_{n+1}(x)p_n(t)$$
$$+ c_{nn}p_n(x)p_n(t) + \frac{a_{n-1}}{a_n} p_{n-1}(x)p_n(t).$$

If this is subtracted from the corresponding identity with t and x interchanged, the term $c_{nn}p_n(t)p_n(x)$ cancels, and the result may be arranged in the form

$$(t - x)p_n(t)p_n(x) = \frac{a_n}{a_{n+1}} \left[p_{n+1}(t)p_n(x) - p_n(t)p_{n+1}(x)\right]$$
$$- \frac{a_{n-1}}{a_n} \left[p_n(t)p_{n-1}(x) - p_{n-1}(t)p_n(x)\right].$$

Let this be written also with n replaced successively by $n-1$, $n-2$, $\cdots$, 0. By addition of the set of $n+1$ relations thus obtained,

$$(t - x) \sum_{k=0}^{n} p_k(t) p_k(x) = \frac{a_n}{a_{n+1}} \left[p_{n+1}(t) p_n(x) - p_n(t) p_{n+1}(x) \right],$$

or with the notation of (6)

$$K_n(x, t) = \frac{a_n}{a_{n+1}} \frac{p_{n+1}(t) p_n(x) - p_n(t) p_{n+1}(x)}{t - x}.$$

This identity, which in the case of Legendre polynomials is essentially that of Christoffel, was generalized by Darboux[*] to systems of orthogonal polynomials with an arbitrary weight function.

7. Symmetry. As a special hypothesis of some interest, let the interval of orthogonality be an interval $(-c, c)$ (or $(-\infty, \infty)$), *symmetric with respect to the origin*, and let $\rho(x)$ be an *even function*. If $q(x)$ is a polynomial, $q(-x)$ is a polynomial of the same degree. Let $q(x)$ denote an arbitrary polynomial of degree lower than the nth. In the integral

$$I = \int_{-c}^{c} \rho(x) p_n(-x) q(x) dx$$

let $t = -x$; then

$$I = \int_{-c}^{c} \rho(-t) p_n(t) q(-t) dt = \int_{-c}^{c} \rho(t) p_n(t) q(-t) dt = 0,$$

since $p_n(t)$ is orthogonal to every polynomial of lower degree with respect to the weight function. The vanishing of I means that $p_n(-x)$ has the same property of orthogonality. Furthermore, $p_n(-x)$ is normalized, and $(-1)^n p_n(-x)$ has a positive coefficient for x^n. Consequently $(-1)^n p_n(-x)$ *is identical with* $p_n(x)$. That is to

[*] See Darboux, pp. 411–414.

say, $p_n(x)$ contains only even powers of x or only odd powers of x, according as n is even or odd.

In particular, the coefficient denoted by b_n in §5 is zero for all values of n in the case of symmetry under discussion, and in this case therefore $c_{nn} = 0$ in (10).

8. Zeros. By an argument which as applied to the special case of Legendre polynomials may be regarded as a supplement to Chapter II, but which is equally simple when formulated for the orthogonal polynomials corresponding to an arbitrary weight function on an arbitrary interval, it can be shown that *the roots of the equation $p_n(x) = 0$ are all real and distinct and interior to the interval* (a, b). If the interval is $(-\infty, \infty)$, the conclusion is merely that the roots are real and distinct.

Since $p_n(x)$ is a polynomial of the nth degree, an equivalent assertion is that $p_n(x)$ changes sign n times in the interior of the interval. In the first place, since $p_n(x)$ is orthogonal to a polynomial of zero degree with respect to the weight function, if $n > 0$,

$$\int_a^b \rho(x) p_n(x) dx = 0.$$

This would certainly not be true if $p_n(x)$ did not change sign in the interval at all. Suppose it changes sign between a and b at just m points $x_1, x_2, \cdots, x_m$. Let

$$\pi(x) = (x - x_1)(x - x_2) \cdots (x - x_m),$$

a polynomial of the mth degree. The product $p_n(x)\,\pi(x)$ does not change sign in the interval:

$$\int_a^b \rho(x) p_n(x) \pi(x) dx \neq 0.$$

If $m < n$, this contradicts the property of orthogonality of $p_n(x)$. It must be therefore that $m = n$.

9. Least-square property. Let $f(x)$ be an arbitrary function, subject to conditions of integrability, and let c_k and $s_n(x)$ be defined by (3) and (4). Let $r_n(x) = f(x) - s_n(x)$. As an immediate consequence of the definitions,

(11)
$$\int_a^b \rho(x) s_n(x) p_k(x) dx = c_k,$$
$$\int_a^b \rho(x) r_n(x) p_k(x) dx = 0, \qquad k = 0, 1, \cdots, n.$$

Let $\pi_n(x)$ be an arbitrary polynomial of the nth degree at most. Let

$$\pi_n(x) - s_n(x) = \delta_n(x) = \sum_{k=0}^{n} d_k p_k(x),$$

so that $\pi_n(x)$ reduces to $s_n(x)$ if the d's are zero, and

$$f(x) - \pi_n(x) = r_n(x) - \delta_n(x).$$

The integral of the square of the error of $\pi_n(x)$ as an approximating function for $f(x)$, weighted from point to point of the interval by the factor $\rho(x)$, is

$$\int_a^b \rho(x) [f(x) - \pi_n(x)]^2 dx$$
$$= \int_a^b \rho(x) [r_n(x) - \delta_n(x)]^2 dx$$
$$= \int_a^b \rho(x) [r_n(x)]^2 dx - 2 \int_a^b \rho(x) r_n(x) \delta_n(x) dx$$
$$+ \int_a^b \rho(x) [\delta_n(x)]^2 dx.$$

But by virtue of (11) and the expression of $\delta_n(x)$ in terms of the p's

$$\int_a^b \rho(x) r_n(x) \delta_n(x) dx = 0, \qquad \int_a^b \rho(x) \big[\delta_n(x)\big]^2 dx = \sum_{k=0}^n d_k^2.$$

Consequently

$$\int_a^b \rho(x) \big[f(x) - \pi_n(x)\big]^2 dx > \int_a^b \rho(x) \big[r_n(x)\big]^2 dx$$

unless all the d's vanish. *The polynomial $s_n(x)$ can be characterized among all polynomials of the nth or lower degree as the one for which the integral of the weighted square of the error is a minimum.*

In particular, let $f(x)$ be the function x^n, and let the c's be the coefficients by means of which x^n is expressed in terms of the p's:

$$x^n = c_0 p_0(x) + c_1 p_1(x) + \cdots + c_n p_n(x).$$

Let the conclusion of the preceding paragraph be applied to the approximation of x^n by polynomials of degree $n - 1$ at most. The polynomial of best approximation in the sense of the least-square criterion is

$$s_{n-1}(x) = c_0 p_0(x) + c_1 p_1(x) + \cdots + c_{n-1} p_{n-1}(x).$$

But

$$x^n - s_{n-1}(x) = c_n p_n(x).$$

The polynomial $p_n(x)$ is determined except for a constant factor by subtracting from x^n the polynomial of degree $n - 1$ at most which makes the integral of the weighted square of the difference a minimum.

10. Differential equation. Let it be supposed now that $\rho(x)$ *satisfies a differential equation*

$$(12) \qquad \frac{\rho'(x)}{\rho(x)} = \frac{D + Ex}{A + Bx + Cx^2},$$

in which A, B, C, D, E are constants, and that $(A + Bx + Cx^2)\rho(x)$ vanishes at the ends of the interval. This is of course a highly restrictive assumption, but the particular weight functions which satisfy it are those of the greatest theoretical and practical importance. The equation (12) is the Pearson differential equation of Chapter VI, which was treated there because of its bearing in the present connection. The argument of this section can be read without reference to the earlier chapter, however, except that in the case of an infinite interval it is to be assumed as a supplementary hypothesis, if not as known otherwise, that the product of $\rho(x)$ by an arbitrary polynomial approaches zero as x becomes infinite in the interval. Under the conditions stated it is to be shown* that $p_n(x)$ *satisfies a differential equation of the form*

$$\alpha(x)p_n''(x) + \beta(x)p_n'(x) + \gamma_n p_n(x) = 0,$$

in which $\alpha(x)$ is a polynomial of the second degree, specifically $A + Bx + Cx^2$, $\beta(x)$ is a polynomial of the first degree independent of n, and γ_n depends on n but is independent of x.

Let $q_m(x)$ be an arbitrary polynomial of degree $m < n$. Let the denominator $A + Bx + Cx^2$ be denoted by $G(x)$. In the integral

$$I = \int_a^b q_m(x) \frac{d}{dx} \left[G(x)\rho(x)p_n'(x) \right] dx$$

---

* See M. Marden, *A rule of signs involving certain orthogonal polynomials*, Annals of Mathematics, (2), vol. 33 (1932), pp. 118–124; p. 120.

let integration by parts be applied with

$$u = q_m(x), \qquad dv = \frac{d}{dx} [G(x)\rho(x)p_n'(x)]dx,$$

$$du = q_m'(x)dx, \qquad v = G(x)\rho(x)p_n'(x).$$

The function uv, being the product of $G(x)\ \rho(x)$ by a polynomial, vanishes at both ends of the interval, and

$$I = - \int_a^b G(x)\rho(x)p_n'(x)q_m'(x)dx.$$

Let another integration by parts be performed, with

$$u = G(x)\rho(x)q_m'(x), \qquad dv = p_n'(x)dx,$$

$$du = \frac{d}{dx}[G(x)\rho(x)q_m'(x)]dx, \qquad v = p_n(x).$$

Again uv vanishes at both ends of the interval, and

$$I = \int_a^b p_n(x)\frac{d}{dx}[G(x)\rho(x)q_m'(x)]dx.$$

But

$$\frac{d}{dx}[G(x)\rho(x)q_m'(x)]$$

$$= (B+2Cx)\rho(x)q_m'(x) + G(x)\rho'(x)q_m'(x) + G(x)\rho(x)q_m''(x)$$

$$= \rho(x)[(B+2Cx)q_m'(x) + (D+Ex)q_m'(x) + G(x)q_m''(x)],$$

since $G(x)\rho'(x) = (D+Ex)\rho(x)$; and the expression in the last bracket is a polynomial $r_m(x)$, of degree m at most. So

$$I = \int_a^b \rho(x)p_n(x)r_m(x)dx = 0,$$

by the property of orthogonality of $p_n(x)$.

In the original expression for I,

$$\frac{d}{dx}\left[G(x)\rho(x)p_n'(x)\right]$$

$$= (B+2Cx)\rho(x)p_n'(x)+G(x)\rho'(x)p_n'(x)+G(x)\rho(x)p_n''(x)$$

$$= \rho(x)\left[(B+2Cx)p_n'(x)+(D+Ex)p_n'(x)+G(x)p_n''(x)\right],$$

which has the form $\rho(x)\,\pi_n(x)$, with a polynomial of the nth degree at most for the second factor. So, for arbitrary $q_m(x)$ of degree $m < n$,

$$\int_a^b \rho(x)\pi_n(x)q_m(x)dx = 0.$$

That is to say, $\pi_n(x)$ has the property of orthogonality which for given n is possessed only by a constant multiple of $p_n(x)$. There must be a constant K_n such that

$$\pi_n(x) = G(x)p_n''(x) + \left[(B+D)+(2C+E)x\right]p_n'(x)$$

$$= K_n p_n(x).$$

By comparison of the coefficients of x^n, the leading coefficient in $p_n(x)$ being denoted once more by $a_n \neq 0$,

$$Cn(n-1)a_n + (2C+E)na_n = K_n a_n,$$

$$K_n = Cn(n+1) + En.$$

So finally

$$(13)\quad \begin{aligned}(A+Bx+Cx^2)p_n''(x)+\left[(B+D)+(2C+E)x\right]p_n'(x)\\ -\left[Cn(n+1)+En\right]p_n(x)=0.\end{aligned}$$

In the case of the Legendre polynomials, $\rho(x)=1$. The hypotheses of the above discussion, including the condition that $G(x)\rho(x)$ shall vanish at the ends of the interval, are satisfied if the differential equation $\rho'(x) = 0$ is written in the form $\rho'(x)/\rho(x) = 0/(1-x^2)$, with

$$B = D = E = 0, \qquad A = 1, \qquad C = -1.$$

Insertion of these constants in (13) reproduces the familiar equation

$$(1 - x^2)p_n''(x) - 2xp_n'(x) + n(n+1)p_n(x) = 0.$$

It was seen in Chapter VI that the Pearson differential equation has essentially three types of solution which give finite integrals over the associated ranges after multiplication by a polynomial of arbitrary degree, the interval being in one case finite, in one case infinite in both directions, and in one case infinite in one direction and bounded in the other. The corresponding systems of orthogonal polynomials, bearing respectively the names of Jacobi, Hermite, and Laguerre, will be studied separately in the next three chapters.

SUPPLEMENTARY REFERENCES: Szegö; Shohat; Pólya-Szegö; Courant-Hilbert; Darboux; Kaczmarz-Steinhaus; Riemann-Weber.

CHAPTER VIII

JACOBI POLYNOMIALS

1. Derivative definition. The domain of orthogonality of the Jacobi polynomials is a finite interval, which may without essential loss of generality be taken as $(-1, 1)$, since an arbitrary finite interval can be reduced to this by a linear change of variable. The weight function is

$$\rho(x) = (1 - x)^\alpha (1 + x)^\beta,$$

in which the exponents are arbitrary real numbers satisfying the condition that $\alpha > -1$, $\beta > -1$, this restriction being imposed so that $\rho(x)$ shall be integrable from -1 to 1. The orthogonal polynomials will be defined outright by means of a derivative formula generalizing the Rodrigues formula for the Legendre polynomials. The property of orthogonality will be deduced from this formula, and the connection with the general theory of Chapter VII will become further apparent as the discussion continues. A similar procedure will be followed in the next two chapters.

For given α, β, n let

$$\phi_n(x) = g(x)h(x), \ g(x) = (1 - x)^{\alpha+n}, \ h(x) = (1 + x)^{\beta+n};$$

the subscript n may be omitted from the auxiliary functions g, h. If α and β are not integers, the indicated powers being then susceptible of interpretation as multiple-valued functions, it is to be understood that those determinations are chosen which make $g(x)$ and $h(x)$ real and positive for $-1 < x < 1$. By Leibniz's formula for the nth derivative of a product,

$$\phi_n^{(n)}(x) = g^{(n)}(x)h(x) + ng^{(n-1)}(x)h'(x)$$
$$+ \frac{n(n-1)}{2!} g^{(n-2)}(x)h''(x) + \cdots + g(x)h^{(n)}(x).$$

For $0 \leqq j \leqq n$, $g^{(j)}(x)$ is a constant multiple of $(1-x)^{\alpha+n-j}$ $= (1-x)^{\alpha}(1-x)^{n-j}$, and $h^{(n-j)}(x)$ is a constant multiple of $(1+x)^{\beta}(1+x)^{j}$; $\phi_n^{(n)}(x)$ has $(1-x)^{\alpha}(1+x)^{\beta}$ as a factor, being the product of this factor by a polynomial of the nth degree at most. Let $y_n(x)$ denote this polynomial:

$$(1) \quad y_n(x) = (1-x)^{-\alpha}(1+x)^{-\beta} \frac{d^n}{dx^n}[(1-x)^{\alpha+n}(1+x)^{\beta+n}].$$

The fact that $y_n(x)$ is actually of the nth and not of lower degree will become apparent presently.

In a standard notation, the *Jacobi polynomial* $P_n^{(\alpha,\beta)}(x)$ is $y_n(x)$ multiplied by $(-1)^n/(2^n n!)$:

$$P_n^{(\alpha,\beta)}(x) =$$
$$\frac{(-1)^n}{2^n n!}(1-x)^{-\alpha}(1+x)^{-\beta}\frac{d^n}{dx^n}[(1-x)^{\alpha+n}(1+x)^{\beta+n}].$$

This is a direct generalization of the Rodrigues formula for the Legendre polynomials, to which it reduces for $\alpha = \beta = 0$.

2. Orthogonality. If $\phi_n^{(k)}(x)$ is expanded by Leibniz's formula for $k < n$, neither of the exponents $\alpha+n$, $\beta+n$ is diminished by more than k units in any term of the expansion, the whole expression has $(1-x)^{\alpha+n-k}(1+x)^{\beta+n-k}$ as a factor, and since

$$\alpha + n - k \geqq \alpha + 1 > 0, \quad \beta + n - k \geqq \beta + 1 > 0,$$

$\phi_n^{(k)}(x)$ vanishes for $x = \pm 1$.

Let m, n be any two non-negative integers, m being the smaller if they are unequal, and let

$$I = \int_{-1}^{1} (1-x)^{\alpha}(1+x)^{\beta} y_m(x) y_n(x) dx$$

$$= \int_{-1}^{1} y_m(x) \phi_n^{(n)}(x) dx.$$

For integration by parts in the second form of the integral let

$$u = y_m(x), \qquad dv = \phi_n^{(n)}(x) dx,$$

$$du = y_m'(x) dx, \qquad v = \phi_n^{(n-1)}(x).$$

Since v vanishes at both ends of the interval, by the preceding paragraph,

$$I = - \int_{-1}^{1} y_m'(x) \phi_n^{(n-1)}(x) dx.$$

By m successive applications of the process,

$$(2) \qquad I = (-1)^m \int_{-1}^{1} y_m^{(m)}(x) \phi_n^{(n-m)}(x) dx.$$

If $m < n$, one more integration by parts gives

$$I = (-1)^{m+1} \int_{-1}^{1} y_m^{(m+1)}(x) \phi_n^{(n-m-1)}(x) dx = 0,$$

since y_m is a polynomial of the mth degree at most, and $y_m^{(m+1)}(x) \equiv 0$. By reference to the original expression for I it appears then that $y_m(x)$ and $y_n(x)$ are orthogonal with respect to the weight function $(1-x)^{\alpha}(1+x)^{\beta}$.

For $m = n$, (2) takes the form

$$I = (-1)^n \int_{-1}^{1} y_n^{(n)}(x)\phi_n(x)dx.$$

Let α_n be the coefficient of x^n in $y_n(x)$ when the terms are collected according to powers of x. Then $y_n^{(n)}(x) = n!\alpha_n$, and

$$\text{(3)} \quad \int_{-1}^{1} (1-x)^{\alpha}(1+x)^{\beta}[y_n(x)]^2 dx$$
$$= I = (-1)^n n!\alpha_n \int_{-1}^{1} \phi_n(x)dx.$$

Since the first integral in (3) is necessarily positive it follows that $\alpha_n \neq 0$. The value of α_n will be determined in the next section, and the evaluation of I in (3) will be further discussed in the section following.

3. Leading coefficients. If α and β are integers the right-hand member of (1) (with the obvious interpretation for $x = \pm 1$) represents the polynomial $y_n(x)$ unambiguously for all values of x, without restriction to the interval $(-1, 1)$. It can be written alternatively in the form

$$\text{(4)} \quad (-1)^n(x-1)^{-\alpha}(x+1)^{-\beta} \frac{d^n}{dx^n}\left[(x-1)^{\alpha+n}(x+1)^{\beta+n}\right].$$

The latter expression represents $y_n(x)$ for $x > 1$ when α and β are not integers, if the powers of $x-1$ and $x+1$ are understood to be real and positive. This is immediately apparent if the Leibniz expansions for the nth derivatives in (1) and (4) are written out explicitly* (without subsequent rearrangement of terms according to powers of x).

---

* The fact can of course be recognized also by tracing the determinations of the multiple-valued function $(1-x)^{\alpha}$ around the branch point $x = 1$ in the complex x-plane.

For $x > 1$ the product

$$(x-1)^{\alpha+n}(x+1)^{\beta+n} = x^{\alpha+\beta+2n}\left(1-\frac{1}{x}\right)^{\alpha+n}\left(1+\frac{1}{x}\right)^{\beta+n}$$

can be expanded by application of the binomial theorem to the last two factors in the form

$$x^{\alpha+\beta+2n} + (\beta - \alpha)x^{\alpha+\beta+2n-1} + \cdots,$$

the terms of positive degree in this expression being followed in the case of non-integral α, β by a series involving negative powers of x. So (for $n > 0$)

$$\frac{d^n}{dx^n}\left[(x-1)^{\alpha+n}(x+1)^{\beta+n}\right] = \lambda_n x^{\alpha+\beta+n} + \mu_n x^{\alpha+\beta+n-1} + \cdots,$$

$$\lambda_n = (\alpha+\beta+2n)(\alpha+\beta+2n-1)\cdots(\alpha+\beta+n+1),$$

$$\mu_n = (\beta-\alpha)(\alpha+\beta+2n-1)(\alpha+\beta+2n-2)\cdots(\alpha+\beta+n),$$

and from the representation (4)

$$y_n(x) = (-1)^n x^{-\alpha-\beta}\left(1-\frac{1}{x}\right)^{-\alpha}\left(1+\frac{1}{x}\right)^{-\beta}(\lambda_n x^{\alpha+\beta+n}$$

$$+ \mu_n x^{\alpha+\beta+n-1} + \cdots)$$

$$= (-1)^n\left(1+\frac{\alpha}{x}+\cdots\right)\left(1-\frac{\beta}{x}+\cdots\right)(\lambda_n x^n$$

$$+ \mu_n x^{n-1} + \cdots)$$

$$= \alpha_n x^n + \beta_n x^{n-1} + \cdots,$$

$$\alpha_n = (-1)^n \lambda_n,$$

$$\beta_n = (-1)^n\left[\mu_n + (\alpha-\beta)\lambda_n\right]$$

$$= (-1)^n(\alpha-\beta)n(\alpha+\beta+2n-1)\cdots(\alpha+\beta+n+1).$$

In more compact representation by means of the

Gamma function, the coefficients of x^n and x^{n-1} in $y_n(x)$ are respectively

(5)
$$\alpha_n = (-1)^n \frac{\Gamma(\alpha + \beta + 2n + 1)}{\Gamma(\alpha + \beta + n + 1)},$$

$$\beta_n = (-1)^n(\alpha - \beta)n \frac{\Gamma(\alpha + \beta + 2n)}{\Gamma(\alpha + \beta + n + 1)}.$$

It will be noted that $\beta_n = 0$ if $\alpha = \beta$, in agreement with §7 of Chapter VII.

For $n = 0$, $y_0(x) \equiv 1$ for all α, β. The formula for α_n in (5) may be regarded as valid even then, except that it becomes meaningless if at the same time $\alpha + \beta + 1 = 0$.

4. Normalizing factor; series of Jacobi polynomials. In the integral

$$J = \int_{-1}^{1} \phi_n(x) dx = \int_{-1}^{1} (1 - x)^{\alpha+n}(1 + x)^{\beta+n} dx$$

let

$$t = \tfrac{1}{2}(1 + x), \qquad 1 - t = \tfrac{1}{2}(1 - x), \qquad dx = 2dt.$$

Then

$$J = 2^{\alpha+\beta+2n+1} \int_0^1 t^{\beta+n}(1 - t)^{\alpha+n} dt$$

$$= 2^{\alpha+\beta+2n+1} B(\beta + n + 1, \alpha + n + 1)$$

$$= 2^{\alpha+\beta+2n+1} \frac{\Gamma(\alpha + n + 1)\Gamma(\beta + n + 1)}{\Gamma(\alpha + \beta + 2n + 2)}.$$

This value of J and the determination of α_n in (5) may now be used for the evaluation of I in (3). If δ_n denotes the corresponding integral with $P_n^{(\alpha,\beta)}(x)$ in place of $y_n(x)$:

$$\delta_n = \int_{-1}^{1} (1 - x)^{\alpha}(1 + x)^{\beta}\left[P_n^{(\alpha,\beta)}(x)\right]^2 dx = \frac{I}{2^{2n}(n!)^2},$$

combination of the various factors gives

$$(6) \quad \delta_n = \frac{2^{\alpha+\beta+1}}{\alpha + \beta + 2n + 1} \; \frac{\Gamma(\alpha + n + 1)\Gamma(\beta + n + 1)}{n!\Gamma(\alpha + \beta + n + 1)}.$$

If $n = \alpha+\beta+1 = 0$, $\delta_0 = \Gamma(\alpha+1)\Gamma(\beta+1)$.

This calculation contributes to the definition of the *normalized Jacobi polynomials* $p_n^{(\alpha,\beta)}(x)$, which are given by

$$p_n^{(\alpha,\beta)}(x) = P_n^{(\alpha,\beta)}(x)/\delta_n^{1/2}.$$

The development of a function $f(x)$ in series of the normalized polynomials is merely a special case under §4 of Chapter VII, with $a = -1$, $b = 1$, and

$$\rho(x) = (1 - x)^{\alpha}(1 + x)^{\beta}.$$

In terms of the polynomials $P_n^{(\alpha,\beta)}(x)$ the formulas become

$$f(x) = \sum_{k=0}^{\infty} C_k P_k^{(\alpha,\beta)}(x), \quad C_k = \frac{1}{\delta_k} \int_{-1}^{1} \rho(x)f(x)P_k^{(\alpha,\beta)}(x)dx.$$

5. Recurrence formula. Let $P_n^{(\alpha,\beta)}(x)$ and $p_n^{(\alpha,\beta)}(x)$ be denoted more simply by $P_n(x)$ and $p_n(x)$. As in §5 of Chapter VII let the coefficients of x^n and x^{n-1} in $p_n(x)$ be denoted by a_n, b_n. In comparison with the coefficients α_n, β_n in §3 of the present chapter, the corresponding coefficients in $P_n(x)$ are, respectively, $(-1)^n\alpha_n/(2^n n!)$ and $(-1)^n\beta_n/(2^n n!)$; and

$$a_n = \frac{(-1)^n\alpha_n}{2^n n!\delta_n^{1/2}}, \qquad b_n = \frac{(-1)^n\beta_n}{2^n n!\delta_n^{1/2}}.$$

From (5),

$$\frac{\alpha_n}{\alpha_{n+1}} = -\frac{\alpha + \beta + n + 1}{(\alpha + \beta + 2n + 1)(\alpha + \beta + 2n + 2)},$$

$$\frac{a_n}{a_{n+1}} = -2(n+1)\left(\frac{\delta_{n+1}}{\delta_n}\right)^{1/2}\frac{\alpha_n}{\alpha_{n+1}}$$

$$= \frac{2(n+1)(\alpha + \beta + n + 1)}{(\alpha + \beta + 2n + 1)(\alpha + \beta + 2n + 2)}\left(\frac{\delta_{n+1}}{\delta_n}\right)^{1/2},$$

$$\frac{b_n}{a_n} = \frac{\beta_n}{\alpha_n} = \frac{(\alpha - \beta)n}{\alpha + \beta + 2n},$$

$$\frac{b_n}{a_n} - \frac{b_{n+1}}{a_{n+1}} = \frac{\beta^2 - \alpha^2}{(\alpha + \beta + 2n)(\alpha + \beta + 2n + 2)}.$$

These values may be substituted in the recurrence formula (10) of Chapter VII. On replacement of $p_k(x)$ by $P_k(x)/\delta_k^{1/2}$ for $k = n+1,\ n,\ n-1$, the formula becomes, after substitution from (6) and simplification,

$$(\alpha + \beta + 2n)(\alpha + \beta + 2n + 1)(\alpha + \beta + 2n + 2)xP_n(x)$$

$$= 2(n+1)(\alpha + \beta + n + 1)(\alpha + \beta + 2n)P_{n+1}(x)$$

$$+ (\beta^2 - \alpha^2)(\alpha + \beta + 2n + 1)P_n(x)$$

$$+ 2(\alpha + n)(\beta + n)(\alpha + \beta + 2n + 2)P_{n-1}(x).$$

This reduces for $\alpha = \beta = 0$ to the familiar recurrence relation for the Legendre polynomials.

6. Differential equation. The weight function $\rho(x) = (1-x)^\alpha(1+x)^\beta$ satisfies the equation

$$\frac{\rho'(x)}{\rho(x)} = -\frac{\alpha}{1-x} + \frac{\beta}{1+x} = \frac{(\beta - \alpha) - (\alpha + \beta)x}{1 - x^2},$$

which has the form of (12) in Chapter VII with

$$A = 1, \quad B = 0, \quad C = -1, \quad D = \beta - \alpha, \quad E = -(\alpha + \beta).$$

The accompanying condition that $(1 - x^2)\rho(x)$ vanish at the ends of the interval is satisfied by virtue of the hypothesis that $\alpha > -1, \beta > -1$. Since the resulting differential equation for the orthogonal polynomials, (13) of Chapter VII, is homogeneous, it is independent of a constant factor in each polynomial, and in particular has the same form whether written for $y_n(x)$ or $p_n(x)$ or $P_n(x)$. In terms of $P_n(x)$ the differential equation reads

$$(1 - x^2)P_n''(x) + [\beta - \alpha - (\alpha + \beta + 2)x]P_n'(x)$$
$$+ n(\alpha + \beta + n + 1)P_n(x) = 0.$$

This reduces to the differential equation of the Legendre polynomials for $\alpha = \beta = 0$, and to equation (10) of Chapter V, except for notation, if α and β have a common positive integral value. In making the latter comparison it is to be noted that the degree of the polynomial $z = P_m^{(n)}(x)$ in the equation of the earlier chapter is neither m nor n, but $m - n$, while the weight function is $(1 - x^2)^n$, so that the present α and β are to be replaced by n and the present n by $m - n$.

For definition of the Jacobi polynomials by means of the coefficients in the power series representation of a "generating function" the reader is referred to more extensive treatises.* The original memoir on the Jacobi polynomials, after derivation of the generating function, remarks:† "This formula, which did not seem to

* See Szegö, pp. 68–70.

† C. G. J. Jacobi, *Untersuchungen über die Differentialgleichung der hypergeometrischen Reihe* (published after Jacobi's death by E.

recommend itself by simplicity, has not been followed up further." Important use has been made of it by subsequent writers,* but it will not be discussed here.†

SUPPLEMENTARY REFERENCES: Szegö; Shohat; Pólya-Szegö; Courant-Hilbert; Riemann-Weber.

Heine), Journal für die reine und angewandte Mathematik, vol. 56 (1859), pp. 149–165; p. 158.

* See e.g. Darboux, pp. 20–24.

† For a generating function in the comparatively simple case in which $\beta = \alpha$ see Ex. 6 in the exercises on this chapter at the end of the book.

CHAPTER IX

HERMITE POLYNOMIALS

1. Derivative definition. The Hermite polynomials are orthogonal over the interval $(-\infty, \infty)$. The weight function* is $e^{-x^2/2}$.

Let

$$\phi(x) = e^{-x^2/2}.$$

By straightforward differentiation,

$$\phi'(x) = -xe^{-x^2/2}, \qquad \phi''(x) = (x^2 - 1)e^{-x^2/2},$$
$$\phi'''(x) = (-x^3 + 3x)e^{-x^2/2}, \cdots.$$

It is clear by inspection, and will presently be apparent on the basis of a more formal induction, that the derivative of any order is the product of $e^{-x^2/2}$ by a polynomial in x.

Let

$$H_n(x) = (-1)^n e^{x^2/2} \frac{d^n}{dx^n} \phi(x).$$

Then $\phi^{(n)}(x) = (-1)^n e^{-x^2/2} H_n(x)$, and by differentiation of this identity

$$\phi^{(n+1)}(x) = (-1)^n \left[-xH_n(x) + H_n'(x) \right] e^{-x^2/2},$$

while on the other hand $\phi^{(n+1)}(x) = (-1)^{n+1} e^{-x^2/2} H_{n+1}(x)$, so that

$$(1) \qquad H_{n+1}(x) = xH_n(x) - H_n'(x).$$

* There is some diversity of usage with regard to the notation. The exponential e^{-x^2} is sometimes taken as weight function instead of $e^{-x^2/2}$; and notations differ otherwise by a factor $(-1)^n$ or a factor $n!$ in the general polynomial of the sequence.

Since $H_0(x) = 1$, it follows by induction that $H_n(x)$ is a polynomial of the nth degree with the coefficient of x^n equal to unity. The polynomials $H_n(x)$ are the *Hermite polynomials*.

2. Orthogonality and normalizing factor. Let m, n have arbitrary non-negative integral values, and let n be the larger if they are unequal: $m \leq n$. Let

$$I = \int_{-\infty}^{\infty} e^{-x^2/2} H_m(x) H_n(x) dx = (-1)^n \int_{-\infty}^{\infty} H_m(x) \phi^{(n)}(x) dx.$$

For an integration by parts, let

$$u = H_m(x), \qquad dv = \phi^{(n)}(x) dx,$$
$$du = H_m'(x) dx, \qquad v = \phi^{(n-1)}(x).$$

Then uv, being the product of $e^{-x^2/2}$ by a polynomial, vanishes for $x = \pm \infty$, and

$$I = (-1)^{n+1} \int_{-\infty}^{\infty} H_m'(x) \phi^{(n-1)}(x) dx.$$

Repetition of the process gives after m steps

$$(2) \qquad I = (-1)^{n+m} \int_{-\infty}^{\infty} H_m^{(m)}(x) \phi^{(n-m)}(x) dx.$$

If $n - m > 0$, one more integration by parts, in which $H_m^{(m+1)}(x) \equiv 0$, since $H_m(x)$ is a polynomial of the mth degree (or straightforward integration of $\phi^{(n-m)}(x)$ with the constant coefficient $H_m^{(m)}(x)$) gives $I = 0$. This means that $H_m(x)$ and $H_n(x)$ are orthogonal over the interval $(-\infty, \infty)$ with respect to the weight function $e^{-x^2/2}$.

For $m = n$, (2) becomes

$$\int_{-\infty}^{\infty} e^{-x^2/2} [H_n(x)]^2 dx = I = \int_{-\infty}^{\infty} H_n^{(n)}(x)\phi(x)dx$$

(3)

$$= n! \int_{-\infty}^{\infty} e^{-x^2/2} dx = n!(2\pi)^{1/2}.$$

The *normalized Hermite polynomials* are given by $H_n(x)/[(n!)^{1/2}(2\pi)^{1/4}]$. The discussion will be continued, however, in terms of the original polynomials $H_n(x)$.

(The evaluation of the last integral in (3), a form of the well-known "probability integral," can be carried out in terms of the Gamma and Beta functions by the relations

$$\pi = \int_{-1}^{1} \frac{dx}{(1-x^2)^{1/2}} = \int_{0}^{1} t^{-1/2}(1-t)^{-1/2}dt$$

$$= B(\tfrac{1}{2}, \tfrac{1}{2}) = \Gamma(\tfrac{1}{2})\Gamma(\tfrac{1}{2})/\Gamma(1) = [\Gamma(\tfrac{1}{2})]^2,$$

$$\frac{1}{2}\int_{-\infty}^{\infty} e^{-x^2/2}dx = \int_{0}^{\infty} e^{-x^2/2}dx = 2^{-1/2}\int_{0}^{\infty} t^{-1/2}e^{-t}dt$$

$$= 2^{-1/2}\Gamma(\tfrac{1}{2}),$$

with the substitution $t=\tfrac{1}{2}(1+x)$ in one case and $t=x^2/2$ in the other; it can of course be accomplished more directly in other ways.)

3. Hermite and Gram–Charlier series. The formal expansion of an arbitrary function $f(x)$ in series of Hermite polynomials is

$$f(x) = \sum_{k=0}^{\infty} c_k H_k(x),$$

(4)

$$c_k = \frac{1}{k!(2\pi)^{1/2}} \int_{-\infty}^{\infty} e^{-x^2/2}f(x)H_k(x)dx,$$

the denominator in the formula for c_k being given by (3).

The Hermite polynomials are used in the theory of statistics for the representation of frequency functions over the interval $(-\infty, \infty)$; but for this purpose a series made up of terms of the form $c_k e^{-x^2/2} H_k(x)$ is more advantageous than a series of the polynomials themselves. Let $F(x)$ be a given function, and let $f(x) = e^{x^2/2} F(x)$. Then (4) becomes

$$F(x) = \sum_{k=0}^{\infty} c_k e^{-x^2/2} H_k(x), \quad c_k = \frac{1}{k!(2\pi)^{1/2}} \int_{-\infty}^{\infty} F(x) H_k(x) dx.$$

A series of this form is called a *Gram-Charlier series*.

4. Recurrence formulas; differential equation. The function $\phi(x)$ satisfies the equation

$$\phi'(x) + x\phi(x) = 0.$$

Differentiation of this gives $\phi''(x) + x\phi'(x) + \phi(x) = 0$; the result of n successive differentiations is

$$\phi^{(n+1)}(x) + x\phi^{(n)}(x) + n\phi^{(n-1)}(x) = 0.$$

By expression of the derivatives here in terms of the Hermite polynomials,

$$[(-1)^{n+1} H_{n+1}(x) + x(-1)^n H_n(x)$$
$$+ n(-1)^{n-1} H_{n-1}(x)] e^{-x^2/2} = 0,$$

or

$$(5) \qquad H_{n+1}(x) - x H_n(x) + n H_{n-1}(x) = 0.$$

This is the relation of recurrence connecting three successive Hermite polynomials.

In terms of the normalized polynomials $h_k(x)$

$= H_k(x)/d_k^{1/2}$, with leading coefficients $a_k = 1/d_k^{1/2}$, where $d_k = k!(2\pi)^{1/2}$, (5) takes the form

$$(n + 1)^{1/2}h_{n+1}(x) - xh_n(x) + n^{1/2}h_{n-1}(x) = 0,$$

in agreement with (10) of Chapter VII.

According to (1), $H_{n+1}(x) - xH_n(x) = -H_n'(x)$. By comparison of this with (5),

$$(6) \qquad H_n'(x) = nH_{n-1}(x).$$

Let $H_{n+1}'(x)$ be calculated by substitution of $n+1$ for n in (6), and again by differentiation of (1); identification of the results leads to the differential equation

$$(7) \qquad H_n''(x) - xH_n'(x) + nH_n(x) = 0.$$

The same equation is obtained by substituting in (13) of Chapter VII the values $D = B = C = 0$, $E = -1$, $A = 1$, corresponding to the differential equation $\rho'(x)/\rho(x) = -x$ satisfied by the weight function $e^{-x^2/2}$; the form of (7) is unaltered if $H_n(x)$ is replaced by the normalized $h_n(x)$.

The identity of the coefficients in (5) and (7), while superficially noteworthy, is of no deep significance, since (5), unlike (7), is changed by the introduction of normalizing factors or other factors depending on n.

Successive Hermite polynomials can be calculated explicitly from (5), when H_0 and H_1 are known. Still more convenient is the use of (6), supplemented by calculation of the constant terms by an induction based on (5):

$$H_{n+1}(0) = -nH_{n-1}(0),$$

$$H_{2k}(0) = (-1)^k \cdot 1 \cdot 3 \cdot 5 \cdots (2k - 1), \qquad H_{2k+1}(0) = 0.$$

The first six polynomials of the sequence are

$$H_0(x) = 1, \qquad H_3(x) = x^3 - 3x,$$
$$H_1(x) = x, \qquad H_4(x) = x^4 - 6x^2 + 3,$$
$$H_2(x) = x^2 - 1, \qquad H_5(x) = x^5 - 10x^3 + 15x.$$

5. Generating function. Let $H(x, t)$ denote the function $e^{xt-(t^2/2)}$. Let its expansion in series of powers of t be written in the form

$$H(x, t) = G_0(x) + G_1(x)t + \frac{1}{2!}G_2(x)t^2 + \frac{1}{3!}G_3(x)t^3 + \cdots.$$

It is seen at once that $G_0(x) = 1$, $G_1(x) = x$. It can also be verified immediately that $H(x, t)$ satisfies the differential equation $\partial H/\partial t = (x - t)H$. If this identity is written in terms of the power series representation, comparison of the coefficients of t^n gives

$$\frac{1}{n!}G_{n+1}(x) = \frac{x}{n!}G_n(x) - \frac{1}{(n-1)!}G_{n-1}(x),$$

that is to say

$$G_{n+1}(x) - xG_n(x) + nG_{n-1}(x) = 0.$$

This is the same as the recurrence relation for the Hermite polynomials. Since $G_0 = H_0$ and $G_1 = H_1$, it follows that $G_2 = H_2$, $G_3 = H_3$, $\cdots$, and by induction that $G_n = H_n$ for all values of n. The Hermite polynomials are related to the *generating function $H(x, t)$* by the identity

$$e^{xt-(t^2/2)} = H_0(x) + H_1(x)t + \frac{1}{2!}H_2(x)t^2$$
$$+ \frac{1}{3!}H_3(x)t^3 + \cdots.$$

6. Wave equation of the linear oscillator. An occurrence of the Hermite polynomials in mathematical physics is in connection with a rudimentary form of the *Schrödinger wave equation* in quantum mechanics. While an adequate discussion either of the mathematical or of the physical significance of the formulas would be impracticable here, the formulas themselves are comparatively simple.

The Schrödinger equation for a single particle in a field of force is of the form

$$(8) \qquad \Delta\Psi - w\Psi = \gamma \frac{\partial\Psi}{\partial t},$$

in which Ψ is a function of the geometrical coordinates and the time t, $\Delta\Psi$ is the expression which forms the left-hand member of Laplace's equation in the number of dimensions in question, the sum of the second partial derivatives of Ψ with respect to the variables of a rectangular coordinate system, w is a function of the coordinates but is independent of t, and γ is a constant.

If a solution is sought in the form $\Psi = uT(t)$, where u is independent of t and T is independent of the coordinates, in accordance with the procedure followed in Chapters IV and V for separation of the variables, $\Delta\Psi = T\Delta u$, and (8) becomes

$$(9) \qquad \frac{\Delta u}{u} - w = \frac{\gamma T'}{T}$$

The constant value common to the two members of (9) being denoted by $-\lambda$, the equation $\gamma T' = -\lambda T$ has for solution a constant multiple of $e^{-(\lambda/\gamma)t}$, and u satisfies the equation

$$(10) \qquad \Delta u + (\lambda - w)u = 0.$$

The physical significance of these equations is restricted

to certain values or sets of values of λ, for which solutions of appropriate form exist.

In one dimension, with $w = cx^2$, c being a constant, these are the equations of the *linear oscillator*.* In this case Δu reduces to d^2u/dx^2. Thus (10) becomes

$$(11) \qquad \frac{d^2u}{dx^2} + (\lambda - cx^2)u = 0.$$

Let $v_n = e^{-x^2/4}H_n(x)$, the functions v with different subscripts being then orthogonal over the interval $(-\infty, \infty)$ with unit weight function. By differentiation of the relation $H_n(x) = e^{x^2/4}v_n$ and substitution in (7) it is found that

$$v_n'' + (n + \tfrac{1}{2} - \tfrac{1}{4}x^2)v_n = 0.$$

Now let this equation be written with τ as independent variable instead of x:

$$(12) \qquad v_n''(\tau) + (n + \tfrac{1}{2} - \tfrac{1}{4}\tau^2)v_n(\tau) = 0,$$

and let $\tau = ax$ with constant a, and

$$(13) \qquad u_n(x) = v_n(\tau) = v_n(ax) = e^{-a^2x^2/4}H_n(ax).$$

Then $d^2u_n/dx^2 = a^2v_n''(\tau)$, and the result of substitution in (12) is

$$\frac{d^2u_n}{dx^2} + [(n + \tfrac{1}{2})a^2 - \tfrac{1}{4}a^4x^2]u_n = 0.$$

So if a is taken equal to $(4c)^{1/4}$, $u_n(x)$ as defined by (13) satisfies (11) for the value $(n + \tfrac{1}{2})a^2$ of λ.

SUPPLEMENTARY REFERENCES: Szegö; Shohat; Rietz; Pólya-Szegö; Courant-Hilbert; Uspensky; Kaczmarz-Steinhaus; Riemann-Weber; for §6: Weyl.

* See Weyl, pp. 54–60.

CHAPTER X

LAGUERRE POLYNOMIALS

1. Derivative definition. The interval of orthogonality of the Laguerre polynomials is $(0, \infty)$. The weight function is

$$\rho(x) = x^\alpha e^{-x},$$

in which, for the sake of integrability, it is assumed that $\alpha > -1$. This is a generalization of the simple exponential weight function e^{-x}, with which the name is sometimes more restrictively associated.

Let

$$\phi_n(x) = x^{\alpha+n} e^{-x}.$$

By application of Leibniz's formula to the nth derivative of this product it is apparent that $\phi_n^{(n)}(x)$ is the product of $x^\alpha e^{-x}$ by a polynomial of the nth degree having $(-1)^n$ as coefficient of x^n. Let this polynomial multiplied by $(-1)^n$ be denoted by $L_n^{(\alpha)}(x)$ or more simply $L_n(x)$:

$$(1) \quad L_n^{(\alpha)}(x) = L_n(x) = (-1)^n x^{-\alpha} e^x \frac{d^n}{dx^n} (x^{\alpha+n} e^{-x}).$$

The formula (1) will be taken as definition* of the *Laguerre polynomials*. The leading coefficient is 1 for each value of n.

2. Orthogonality; normalizing factor; Laguerre series. Each derivative of $\phi_n(x)$ of order lower than the

* Notations vary in different presentations to the extent of a constant factor in each polynomial.

nth is the product of $x^{\alpha+1}e^{-x}$ by a polynomial, and vanishes for $x = 0$, since $\alpha > -1$, and for $x = +\infty$.

Let m and n be positive integers, $m \leqq n$, and let

$$I = \int_0^\infty x^\alpha e^{-x} L_m(x) L_n(x) dx = (-1)^n \int_0^\infty L_m(x) \phi_n^{(n)}(x) dx.$$

In m successive integrations by parts, with $u = (d^{k-1}/dx^{k-1})L_m(x)$, $dv = \phi_n^{(n-k+1)}(x)dx$ as factors for the kth step, the expression uv with $v = \phi_n^{(n-k)}(x)$ vanishes at both ends of the interval in each case, and it is found that*

$$(2) \qquad I = (-1)^{n+m} \int_0^\infty (L_m(x))^{(m)} \phi_n^{(n-m)}(x) dx.$$

If $n > m$, another integration by parts leads to the conclusion that $I = 0$, since $(L_m(x))^{(m+1)} \equiv 0$. *The polynomials $L_m(x)$, $L_n(x)$ are orthogonal for weight $x^\alpha e^{-x}$.*

For $m = n$, (2) reduces to

$$I = \int_0^\infty (L_n(x))^{(n)} \phi_n(x) dx.$$

Since the leading coefficient in $L_n(x)$ is 1, $(L_n(x))^{(n)} = n!$. If d_n denotes the value of I in this case,

$$\int_0^\infty x^\alpha e^{-x} [L_n(x)]^2 dx = d_n = n! \int_0^\infty x^{\alpha+n} e^{-x} dx$$
$$= n! \Gamma(\alpha + n + 1).$$

The *normalized Laguerre polynomial* $l_n(x)$ is given by

$$l_n(x) = L_n(x)/d_n^{1/2}.$$

* The notation $(L_m(x))^{(m)}$ is used to avoid confusion with the polynomial $L_m^{(\alpha)}(x)$ corresponding to exponent $\alpha = m$.

The expansion of a function $f(x)$ in series of the polynomials $L_k(x)$ on the interval $(0, \infty)$ is

$$f(x) = \sum_{k=0}^{\infty} c_k L_k(x),$$

$$c_k = \frac{1}{k!\Gamma(\alpha + k + 1)} \int_0^{\infty} x^\alpha e^{-x} f(x) L_k(x) dx.$$

3. Differential equation and recurrence formulas.
For the weight function $\rho(x) = x^\alpha e^{-x}$, since $\rho'(x)/\rho(x) = (\alpha/x) - 1$, equation (12) of Chapter VII is satisfied with

$$D = \alpha, \qquad E = -1, \qquad A = C = 0, \qquad B = 1;$$

and $x\rho(x)$ vanishes at both ends of the interval $(0, \infty)$. So the Laguerre polynomials satisfy (13) of Chapter VII with these values inserted. It is immaterial whether the polynomials are normalized or not; written for $L_n(x)$, the differential equation is

$$(3) \qquad xL_n''(x) + (\alpha + 1 - x)L_n'(x) + nL_n(x) = 0.$$

If a_n denotes the leading coefficient in the normalized polynomial $l_n(x)$, $a_n = 1/d_n^{1/2}$. The coefficient of x^{n-1} in $L_n(x)$, by application of Leibniz's formula in (1), is $-n(\alpha+n)$. So the corresponding coefficient b_n in $l_n(x)$ is $-n(\alpha+n)/d_n^{1/2}$. For substitution in (10) of Chapter VII, $c_{nn} = (b_n/a_n) - (b_{n+1}/a_{n+1}) = \alpha + 2n + 1$. If at the same time $L_n(x)/d_n^{1/2}$ is substituted for $l_n(x)$, the recurrence formula becomes, after rearrangement,

$$(4) \qquad L_{n+1}(x) - (x - \alpha - 2n - 1)L_n(x) + n(\alpha+n)L_{n-1}(x) = 0.$$

Another form of recurrence relation, expressing $L_{n+1}(x)$ in terms of $L_n(x)$ and $L_n'(x)$, is almost immediately deducible from the defining formula (1). In the identity

$$(-1)^{n+1}x^\alpha e^{-x}L_{n+1}(x) = \frac{d^{n+1}}{dx^{n+1}}(x^{\alpha+n+1}e^{-x})$$

let the expression in parentheses on the right be regarded as the product of the factors x and $x^{\alpha+n}e^{-x}$, and let Leibniz's formula be applied to the differentiation of this product. All derivatives of the factor x of order higher than the first are of course identically zero. So

$$\frac{d^{n+1}}{dx^{n+1}}(x^{\alpha+n+1}e^{-x})$$

$$= x\frac{d^{n+1}}{dx^{n+1}}(x^{\alpha+n}e^{-x}) + (n+1)\frac{d^n}{dx^n}(x^{\alpha+n}e^{-x})$$

$$= x\frac{d}{dx}[(-1)^n x^\alpha e^{-x}L_n(x)] + (n+1)(-1)^n x^\alpha e^{-x}L_n(x)$$

$$= (-1)^n x^\alpha e^{-x}[xL_n'(x) + (\alpha+n+1-x)L_n(x)].$$

It follows that

$$L_{n+1}(x) = (x-\alpha-n-1)L_n(x) - xL_n'(x).$$

4. Generating function. Let

$$H(x,t) = \frac{1}{(1-t)^{\alpha+1}}e^{-xt/(1-t)},$$

and in the expansion of this function as a power series in t let the coefficient of t^n be denoted by $(-1)^n G_n(x)/n!$, so that

$$\text{(5)}\qquad \begin{aligned} H(x,t) &= G_0(x) - G_1(x)t + \frac{1}{2!}G_2(x)t^2 + \cdots \\ &\quad + \frac{(-1)^n}{n!}G_n(x)t^n + \cdots. \end{aligned}$$

It is to be shown that $G_n(x)$ *is identical with* $L_n(x)$.

The first two polynomials in the Laguerre sequence are by substitution of $n = 0$ and $n = 1$ in (1)

$$L_0(x) = 1, \qquad L_1(x) = x - \alpha - 1.$$

The result of partial differentiation of H with respect to t is

$$(6) \qquad \frac{\partial}{\partial t} H(x, t) = \left[\frac{\alpha + 1}{1 - t} - \frac{x}{(1 - t)^2} \right] H(x, t).$$

Hence

$$G_0(x) = H(x, 0) = 1, \ \ G_1(x) = -H_t(x, 0) = x - \alpha - 1.$$

It is seen that G_0 and G_1 agree with L_0 and L_1.

Let (6) be written in the form

$$(1 - t)^2 H_t(x, t) + [(x - \alpha - 1) + (\alpha + 1)t] H(x, t) = 0,$$

and let the representation (5) be substituted in the left-hand member of this identity. By equating to zero the coefficient of t^n in the resulting series it is found that

$$G_{n+1} + (2n - x + \alpha + 1)G_n + (n^2 + n\alpha)G_{n-1} = 0.$$

Since this is the same as the relation connecting the polynomials L_{n-1}, L_n, L_{n+1} in (4), and since $G_0 = L_0$, $G_1 = L_1$, it follows by substitution of $n = 1, 2, \cdots$ that $G_n(x) = L_n(x)$ for all values of n.

5. Wave equation of the hydrogen atom. The discussion of the Schrödinger wave equation which was begun in the second and third paragraphs of §6 in Chapter IX becomes applicable to the electron of a hydrogen atom in space if continued with reference to three dimensions instead of one, the function w being suitably chosen. Solutions are obtained then involving Laguerre polynomials.

For this case w is to be taken in the form $(-c/\rho)$,

where ρ is distance from the origin and c is constant.*
The differential equation (10) of Chapter IX becomes

$$(7) \qquad \Delta u + \left(\lambda + \frac{c}{\rho}\right) u = 0.$$

Because of the form of w the use of spherical coordinates
is appropriate, and Δu is given by (26) of Chapter IV.
Then (4) of Chapter V becomes equivalent to the pres-
ent equation (7) if $(\lambda\rho^2 + c\rho)u$ is added to the left-hand
member. On substitution of $R(\rho)T(\theta)F(\phi)$ for u an equation is obtained which differs from (5) of Chapter V by
the addition of $\lambda\rho^2 + c\rho$ on the left. Thus (6) of Chapter
V is replaced by

$$(8) \quad \rho^2 R'' + 2\rho R' + [\lambda\rho^2 + c\rho - m(m+1)]R = 0.$$

The equations in terms of θ and ϕ are unchanged, and
particular functions $T(\theta)F(\phi)$ are obtained as in the
earlier chapter in the form of spherical harmonics. The
problem that remains for discussion is the solution of the
present equation (8). (The factor $T(\theta)$ has no connection
of course with the function $T(t)$ in §6 of Chapter IX;
and the n in (7) and (8) of Chapter V, which there will
be no occasion to mention again explicitly, is not the
same as the n which appears in the next paragraphs.)

In terms of the value of m which occurs in (8) let
$\alpha = 2m + 1$, and let $L_n(x)$ be the corresponding Laguerre
polynomial $L_n^{(\alpha)}(x) = L_n^{(2m+1)}(x)$. By (3),

$$(9) \quad xL_n''(x) + (2m + 2 - x)L_n'(x) + nL_n(x) = 0.$$

Let $v_n(x) = x^m e^{-x/2} L_n(x)$; two v's with different subscripts
are orthogonal on the interval $(0, \infty)$ with respect to x
as weight function. Differentiation of the identity $L_n(x)$
$= x^{-m} e^{x/2} v_n$ gives

* See Courant-Hilbert, pp. 294–296; in connection with this section see also van der Waerden, pp. 12–16, and Weyl, pp. 63–70.

$$L_n'(x) = x^{-m}e^{x/2}\left\{v_n' + \left(\frac{1}{2} - \frac{m}{x}\right)v_n\right\},$$

$$L_n''(x) = x^{-m}e^{x/2}\left\{v_n'' + \left(1 - \frac{2m}{x}\right)v_n'\right.$$

$$\left. + \left[\frac{m}{x^2} + \left(\frac{1}{2} - \frac{m}{x}\right)^2\right]v_n\right\}.$$

By substitution of these expressions in (9) and multiplication of the resulting equation by $x^{m+1}e^{-x/2}$ it is found that

(10) $x^2v_n'' + 2xv_n' + \left[-\frac{1}{4}x^2 + (m+n+1)x - m(m+1)\right]v_n = 0.$

Now let $x = a_n\rho$, where a_n is constant for fixed n but will presently be made to depend on n in a specified manner, and ρ has the same meaning as in the second paragraph preceding; x then is not one of the rectangular coordinates in the physical problem, but is merely an auxiliary variable for the purposes of the present calculation. Let

$$R(\rho) = v_n(x) = v_n(a_n\rho) = (a_n\rho)^m e^{-a_n\rho/2}L_n^{(2m+1)}(a_n\rho).$$

Then $v_n'(x) = (1/a_n)R'(\rho)$, $v_n''(x) = (1/a_n^2)R''(\rho)$, if the accents in each case indicate differentiation with respect to the variable following in parentheses, and substitution in (10) gives

$$\rho^2R'' + 2\rho R' + \left[-\frac{1}{4}a_n^2\rho^2 + (m+n+1)a_n\rho - m(m+1)\right]R = 0.$$

So if a_n is taken equal to $c/(m+n+1)$, the function $R(\rho)$ as now defined satisfies (8) with $\lambda = -a_n^2/4$.

SUPPLEMENTARY REFERENCES: Szegö; Shohat; Pólya-Szegö; Courant-Hilbert; Uspensky; Kaczmarz-Steinhaus; Riemann-Weber; for §5: Courant-Hilbert; van der Waerden; Weyl.

CHAPTER XI

CONVERGENCE

1. Scope of the discussion. The question of the convergence of series of orthogonal polynomials under the most general hypotheses, as might be anticipated, is one of great complexity. A considerable part of the theory of convergence developed for Fourier series in the first chapter, or for Legendre series in the second, can however be extended with little additional difficulty to orthogonal polynomials corresponding to various weight functions on a finite interval *if the polynomials of the orthonormal set are bounded as n becomes infinite*, i.e., if there is a constant H, independent of n, such that $|p_n(x)| \leqq H$ for all values of n, either at a particular point x where convergence is to be demonstrated or on a set of points contained in the interval of orthogonality or, for the greatest simplicity, throughout the entire interval.

The orthonormal polynomials corresponding to the weight function $(1-x^2)^{-1/2}$, for example, are bounded throughout the interval $(-1, 1)$. For if $C_n(x)$ is the polynomial which represents $\cos n\theta$ when $x = \cos \theta$ the sequence of these polynomials for $n = 0, 1, 2, \cdots$ constitutes the orthogonal system, except for normalization (see Chapter VII, §1): $|C_n(x)| = |\cos n\theta| \leqq 1$ for $-1 \leqq x \leqq 1$, and the normalizing factor $\delta_n^{-1/2}$ defined by

$$\delta_n = \int_{-1}^{1} (1 - x^2)^{-1/2} [C_n(x)]^2 dx = \int_{0}^{\pi} \cos^2 n\theta \, d\theta$$

has the constant value $(2/\pi)^{1/2}$ for $n \geqq 1$.

The polynomials $S_n(x)$ of Chapter VII, §1, which are orthogonal but not normalized for weight $(1-x^2)^{1/2}$, are bounded throughout any interval $(-1+h, 1-h)$, $h>0$, that is to say, except near the ends of the interval of orthogonality. For if $x = \cos \theta$, $1-h = \cos \gamma$, $0 < \gamma < \pi/2$, then

$$\left| S_n(x) \right| = \left| \sin (n + 1)\theta / \sin \theta \right| \leq 1/\sin \theta \leq 1/\sin \gamma$$

for $\gamma \leq \theta \leq \pi - \gamma$, $-1+h \leq x \leq 1-h$. And the normalizing factor $1/\delta_n^{1/2}$ is again independent of n, being given by

$$\delta_n = \int_{-1}^{1} (1-x^2)^{1/2} [S_n(x)]^2 dx = \int_{0}^{\pi} \sin^2 (n+1)\theta \ d\theta = \pi/2.$$

An alternative verbal statement is that *the orthonormal polynomials $S_n/\delta_n^{1/2}$ are uniformly bounded throughout any closed interval interior to the interval of orthogonality*; it is clear that any such interval, even if not itself symmetric with respect to the origin, can be included in an interval $(-1+h, 1-h)$ by suitable choice of h.

By §12 of Chapter II, the normalized Legendre polynomials $[(2n+1)/2]^{1/2} P_n(x)$ are uniformly bounded throughout any closed interval interior to $(-1, 1)$.

It will be shown in later sections that the property of being uniformly bounded throughout the interval, or except near the ends of the interval, or over a part of the interval restricted in some other way, is possessed by the orthonormal polynomials corresponding to weight functions of considerable generality. Meanwhile the bearing of this property on the theory of convergence will be developed in some detail.

2. Magnitude of the coefficients; first hypothesis.
Let $\rho(x)$ be a weight function as in Chapter VII, *for a*

finite * *interval* (a, b), let $p_n(x)$, $n = 0, 1, 2, \cdots$, be the
corresponding system of normalized orthogonal polyno-
mials, let $f(x)$ be a function such that $\rho(x)f(x)$ is integrable over (a, b), and let

(1) $$\sum_{k=0}^{\infty} c_k p_k(x), \qquad c_k = \int_a^b \rho(x)f(x)p_k(x)dx,$$

be the formal development of $f(x)$ in series of the polynomials $p_k(x)$. Let $s_n(x)$ be the partial sum of the series
(1) through terms of the nth degree:

$$s_n(x) = \sum_{k=0}^{n} c_k p_k(x).$$

Then, by the definition of the c's and the properties of
the orthonormal p's,

$$\int_a^b \rho(x)f(x)s_n(x)dx = \sum_{k=0}^{n} c_k \int_a^b \rho(x)f(x)p_k(x)dx = \sum_{k=0}^{n} c_k^2,$$

$$\int_a^b \rho(x)\left[s_n(x)\right]^2 dx = \sum_{k=0}^{n} c_k^2.$$

If ρf^2 as well as ρf is integrable,

$$\int_a^b \rho(x)\left[f(x) - s_n(x)\right]^2 dx$$

$$= \int_a^b \rho(x)\left[f(x)\right]^2 dx - 2\int_a^b \rho(x)f(x)s_n(x)dx$$

$$+ \int_a^b \rho(x)\left[s_n(x)\right]^2 dx$$

$$= \int_a^b \rho(x)\left[f(x)\right]^2 dx - \sum_{k=0}^{n} c_k^2.$$

* This restriction is not essentially used until the next section.

Since the first member is non-negative, the same is true of the last member:

$$\sum_{k=0}^{n} c_k^2 \leqq \int_a^b \rho(x) \left[f(x) \right]^2 dx$$

for any value of n. Consequently $\sum c_k^2$ is convergent, and

$$\lim_{k \to \infty} c_k = 0.$$

3. Convergence; first hypothesis. By §§4 and 6 of Chapter VII,

$$(2) \qquad s_n(x) = \int_a^b \rho(t) f(t) K_n(x, t) dt,$$

$$(3) \qquad K_n(x, t) = \frac{a_n}{a_{n+1}} \frac{p_{n+1}(t) p_n(x) - p_n(t) p_{n+1}(x)}{t - x},$$

where, in the notation of §5 of that chapter,

$$(4) \qquad \frac{a_n}{a_{n+1}} = c_{n,n+1} = \int_a^b \rho(x) x p_n(x) p_{n+1}(x) dx.$$

Furthermore, by (7) of the same chapter, with $\pi_n(x) \equiv 1$,

$$1 = \int_a^b \rho(t) K_n(x, t) dt.$$

Multiplication of this by $f(x)$, which is constant with respect to the variable of integration, gives

$$(5) \qquad f(x) = \int_a^b \rho(t) f(x) K_n(x, t) dt.$$

Hence, by subtraction of (5) from (2),

$$(6) \quad s_n(x) - f(x) = \int_a^b \rho(t) \left[f(t) - f(x) \right] K_n(x, t) dt.$$

For fixed x let

$$(7) \qquad \frac{f(t) - f(x)}{t - x} = \phi(t).$$

Then by substitution of (3) in (6), $s_n(x) - f(x)$ is equal to

$$(8) \qquad \frac{a_n}{a_{n+1}} \left[p_n(x) \int_a^b \rho(t)\phi(t)p_{n+1}(t)dt \right.$$
$$\left. - p_{n+1}(x) \int_a^b \rho(t)\phi(t)p_n(t)dt \right].$$

The proof of convergence consists in showing that under suitable hypotheses the value of the expression (8) approaches zero as n becomes infinite.

Let C be the larger of the numbers $|a|$, $|b|$. Then $|x| \leqq C$ throughout (a, b), and according to (4)

$$\frac{a_n}{a_{n+1}} \leqq C \int_a^b \rho(x) \left| p_n(x)p_{n+1}(x) \right| dx;$$

since a_n and a_{n+1} are both positive, absolute value signs on the left would be superfluous. By Schwarz's inequality (see Chapter I, §19)

$$\left\{ \int_a^b [\rho(x)]^{1/2} \left| p_n(x) \right| \cdot [\rho(x)]^{1/2} \left| p_{n+1}(x) \right| dx \right\}^2$$
$$\leqq \int_a^b \rho(x) [p_n(x)]^2 dx \int_a^b \rho(x) [p_{n+1}(x)]^2 dx = 1.$$

So $a_n/a_{n+1} \leqq C$ for all values of n.

If the function $f(t)$ is such that $\rho\phi$ and $\rho\phi^2$ are integrable over (a, b), where $\phi(t)$ is the difference quotient (7) formed for the value of x in question, each of the integ-

rals in (8) approaches zero as n becomes infinite, by the preceding section. *If at the same time $p_n(x)$ is bounded as n becomes infinite,* for the particular value of x concerned, the whole expression (8) approaches zero, and $s_n(x)$ *converges to the value $f(x)$.*

The weight function ρ is of course assumed throughout to be integrable. If integrals are thought of as taken in the ordinary sense of elementary analysis (not according to the more general definition of Lebesgue), let ρ be supposed to have at most a finite number of finite or infinite discontinuities. The above hypothesis with regard to ϕ will be fulfilled, for example, if $f(t)$ is bounded and integrable over the interval, is continuous at the point x, and either has a right-hand and a left-hand derivative there (with finite values) or, somewhat more generally, satisfies the condition that $\left| f(t) - f(x) \right| \leqq \lambda \left| t - x \right|$, where λ is a constant.

The hypothesis concerning ϕ will also be satisfied if ρf and ρf^2 are integrable over the interval, and if $f(t)$ vanishes identically throughout a subinterval, no matter how small, containing the point x in its interior. In this case $s_n(x)$ converges to the value 0. If f_1 and f_2 are two functions subject to the conditions of integrability just stated, and identically equal to each other in the subinterval, the partial sum of the series for the difference $f_1 - f_2$, equal to the difference of the corresponding partial sums, will approach zero; if the series for one of the functions is convergent at the point x, the series for the other will converge to the same value. *If the function f is such that ρf and ρf^2 are integrable over (a, b), the convergence of the series for $f(x)$ at a point where the orthonormal polynomials are bounded is further dependent only on the behavior of the function in the immediate neighborhood of the point.*

4. Magnitude of the coefficients; second hypothesis.*

The conclusion of §2 holds without the requirement that ρf^2 be integrable, if the polynomials $p_n(x)$ are uniformly bounded throughout the interval. More generally, suppose there is a point set E contained in (a, b) with a constant H, independent of n and x, such that $\left| p_n(x) \right| \leqq H$ for all values of n and for all values of x belonging to E. Let CE denote the rest of the interval. In the applications that are of primary interest, E will be merely a subinterval of (a, b), or possibly a set of a finite number of such subintervals. It may be assumed for the purposes of this discussion that E is made up of a finite number of intervals at most, except that if integration is understood in the general sense of Lebesgue the hypothesis is merely that E is measurable (as it will be automatically if it denotes the set of *all* points where $\left| p_n(x) \right| \leqq H$). If ρf and $\rho \left| f \right|$ are integrable over the whole of (a, b), and ρf^2 is integrable over CE, *it is still true that* $\lim_{n \to \infty} c_n = 0$. Under the Lebesgue definition a separate hypothesis with regard to $\rho \left| f \right|$ is unnecessary, since integrability and absolute integrability are the same.

Let ϵ be an arbitrarily small positive number. Let the point set E be resolved into two parts E_1, E_2 such that

$$(9) \qquad \int_{E_1} \rho(x) \left| f(x) \right| dx \leqq \frac{\epsilon}{2H},$$

and $\left| f(x) \right|$ is bounded on E_2, i.e. there is a constant M such that $\left| f(x) \right| \leqq M$ at all points of E_2. In the case of Lebesgue integration E_1 may be defined as the set of all points of E where $\left| f(x) \right| > M$, for a sufficiently large

---

* The content of §§4 and 5, included for the sake of additional generality, is not needed for the reading of the later sections.

value of M, and E_2 as the set of points of E where $|f(x)| \leq M$. For more elementary treatment the purpose can be accomplished by inclosing the points of infinite discontinuity of $f(x)$, supposed finite in number, in sufficiently narrow intervals, and letting E_1 be the part of E contained in these intervals, and E_2 the rest of E.

Let I_1, I_2, I_3 denote the integrals of $\rho(x)f(x)p_n(x)$ over E_1, E_2, and CE respectively, so that

$$\int_a^b \rho(x)f(x)p_n(x)dx = I_1 + I_2 + I_3.$$

By (9),

$$|I_1| \leq \int_{E_1} \rho(x) |f(x)| |p_n(x)| dx$$

(10)

$$\leq H \int_{E_1} \rho(x) |f(x)| dx \leq \frac{\epsilon}{2},$$

regardless of the value of n, since $|p_n(x)| \leq H$ on E_1 for all n. Let $f_1(x)$ be a function defined as identically zero on E_1, and equal to $f(x)$ on the rest of (a, b). Then

$$I_2 + I_3 = \int_a^b \rho(x)f_1(x)p_n(x)dx.$$

But ρf_1^2 is integrable over (a, b); it is integrable over CE by hypothesis, is integrable over E_2 because $|f_1| = |f| \leq M$ on E_2, and hence $\rho f_1^2 \leq M\rho|f|$, and is integrable over E_1 because it is identically zero there. Consequently, by §2,

$$\lim_{n \to \infty} \int_a^b \rho(x)f_1(x)p_n(x)dx = 0,$$

i.e. $\lim_{n \to \infty} (I_2 + I_3) = 0$. So when n is sufficiently large $|I_2 + I_3| < \epsilon/2$, and by combination of this with (10)

$$\left| I_1 + I_2 + I_3 \right| < \epsilon,$$

which means that

$$\lim_{n\to\infty} \int_a^b \rho(x)f(x)p_n(x)dx = 0.$$

5. Convergence; second hypothesis. Let the discussion of convergence be resumed with the notation of §3, under the hypothesis now that the orthonormal polynomials are uniformly bounded on a point set E as in §4. Let it be assumed that ρf and $\rho|f|$ are integrable over (a, b), and that ρf^2 is integrable over the set CE.

If x is an interior point of E, $\rho\phi^2$ will be integrable over CE, the denominator of ϕ having a positive lower bound there; and $\rho\phi$ will be integrable if $\rho|\phi|$ is integrable, so that a separate hypothesis with regard to the former is unnecessary. Furthermore, $\rho|\phi|$ will be integrable over the whole of (a, b) if there is an interval $(x-h, x+h)$ over which it is integrable. *If ρf and $\rho|f|$ are integrable over (a, b) and ρf^2 is integrable over CE, the series will converge to the value $f(x)$ at an interior point x of E, if $\rho|\phi|$ is integrable over an interval containing x in its interior.* For under these conditions ϕ satisfies the requirements imposed on f in §4, and the reasoning of the first three paragraphs of §3 can be repeated with the modification merely that §4 is used instead of §2 to justify the assertion that the integrals in (8) approach zero. The essential difference between the present hypothesis concerning f and that of §3 is that $\rho\phi^2$ is not required to be integrable in the neighborhood of x.

The reasoning applies at the end points a, b, with obvious modifications in details of statement, if E extends out to those points. Special consideration of other frontier points of E may be omitted, since in the applications, when a set E is shown to be associated with a particular value of H, its frontier points (other than a or b) will be interior to the corresponding set associated with a larger value of H.

If the orthonormal polynomials are uniformly bounded throughout the whole of (a, b) no statement about ρf^2 or $\rho \phi^2$ is needed in the hypothesis, and the convergence proof applies throughout the closed interval.

If ρf and $\rho |f|$ are integrable over (a, b) and if ρf^2 is integrable over CE (the last clause being omitted if E is the whole interval), convergence at an interior point of E is independent of the properties of f except in the immediate neighborhood of the point.

6. Special Jacobi polynomials. Let $p_k(x)$ denote the normalized Legendre polynomial of the kth degree. For a pair of even values $2m$, $2n$ of k, let

$$I = \int_{-1}^{1} p_{2m}(x) p_{2n}(x) dx = 2 \int_{0}^{1} p_{2m}(x) p_{2n}(x) dx.$$

In the second integral let

$$x^2 = \tfrac{1}{2}(1 + t), \qquad dx = \frac{dt}{4x} = \frac{1}{2^{3/2}} \frac{dt}{(1 + t)^{1/2}} \cdot$$

Since $p_{2m}(x)$, $p_{2n}(x)$ contain only even powers of x, they are polynomials in t, of degrees m and n respectively, which may be denoted by $Q_m(t)$, $Q_n(t)$, and

$$I = \frac{1}{2^{1/2}} \int_{-1}^{1} Q_m(t) Q_n(t) \frac{dt}{(1 + t)^{1/2}} \cdot$$

If $m \neq n$, $I = 0$. If $m = n$, since the p's are normalized,

$$1 = I = \frac{1}{2^{1/2}} \int_{-1}^{1} [Q_n(t)]^2 \frac{dt}{(1+t)^{1/2}},$$

and the value of the last integral is $2^{1/2}$. *The polynomials* $q_k(t) = Q_k(t)/2^{1/4}$, $k = 0, 1, 2, \cdots$, *constitute the orthonormal system on the interval* $(-1, 1)$ *for weight* $(1+t)^{-1/2}$. They are the normalized Jacobi polynomials for $\alpha = 0$, $\beta = -\frac{1}{2}$. Since the p's are uniformly bounded except when x^2 is near 1, *the q's are uniformly bounded except near the point* $t = 1$, i.e. are uniformly bounded throughout any interval $-1 \leq t \leq 1-h$, $h > 0$.

Similar reasoning with the substitution $x^2 = \frac{1}{2}(1-t)$ in place of that used above shows that the normalized Jacobi polynomials with $\alpha = -\frac{1}{2}$, $\beta = 0$ are uniformly bounded in $(-1, 1)$ except near the point $t = -1$.

These facts, together with those already noted for the Legendre polynomials and the cosine polynomials (see §1), mean that *the normalized Jacobi polynomials are uniformly bounded throughout any closed interval interior to* $(-1, 1)$, *if the exponents* (α, β) *have any of the pairs of values* $(-\frac{1}{2}, -\frac{1}{2})$, $(-\frac{1}{2}, 0)$, $(0, -\frac{1}{2})$, $(0, 0)$.

Combined with a special case of a result to be obtained in the next section, this observation will make it apparent that *a similar conclusion holds for any pair of exponents* (α, β), *each of which is an integer or half of an integer* (with the understanding, of course, that $\alpha > -1$, $\beta > -1$). The above theory of convergence is applicable accordingly. Proof of the corresponding fact for completely arbitrary Jacobi polynomials is outside the scope of this presentation.

7. Multiplication or division of the weight function by a polynomial. Let $\Pi(x)$ be a polynomial which is nonnegative on the interval (a, b). This polynomial, once chosen, is to keep its identity unchanged through the next stages of the discussion. Let its degree be m. Let $p_k(x)$, $k = 0, 1, \cdots$, be the orthonormal polynomials corresponding to a weight function $\rho(x)$ on (a, b), and let $q_k(x)$, $k = 0, 1, \cdots$, be the orthonormal system associated with $\Pi(x)\rho(x)$ as weight function. It is to be shown that properties of boundedness of the polynomials $q_k(x)$ can be inferred from boundedness of the p's.

The product $\Pi(x)q_n(x)$, being a polynomial of degree $n + m$, can be expressed in the form

$$\Pi(x)q_n(x) = \sum_{k=0}^{n+m} c_{nk} p_k(x),$$

$$c_{nk} = \int_a^b \rho(x)\Pi(x)q_n(x)p_k(x)dx.$$

If $k < n$, $c_{nk} = 0$, in consequence of the properties of orthogonality of $q_n(x)$ with respect to the weight function $\Pi(x)\rho(x)$. So

$$(11) \qquad \Pi(x)q_n(x) = \sum_{k=n}^{n+m} c_{nk} p_k(x),$$

with just $m + 1$ terms on the right, instead of a number of terms increasing with n. As to the coefficients which do not vanish,

$$\left| c_{nk} \right| \leqq \int_a^b \left[\rho(x) \right]^{1/2} \Pi(x) \left| q_n(x) \right| \cdot \left[\rho(x) \right]^{1/2} \left| p_k(x) \right| dx,$$

$$c_{nk}^2 \leqq \int_a^b \rho(x) \left[\Pi(x) \right]^2 \left[q_n(x) \right]^2 dx \int_a^b \rho(x) \left[p_k(x) \right]^2 dx,$$

by Schwarz's inequality. The last integral is equal to 1, since the p's are normalized. Let G be the maximum of $\Pi(x)$ on (a, b). Then

$$\int_a^b \rho(x)\left[\,\Pi(x)\,\right]^2[q_n(x)\,]^2dx$$

$$\leq G\int_a^b \rho(x)\,\Pi(x)\,[q_n(x)\,]^2dx = G,$$

since the q's are normalized for weight $\Pi(x)\rho(x)$. So $|c_{nk}| \leq G^{1/2}$.

Let the polynomials $p_k(x)$ satisfy the condition that $|p_k(x)| \leq H$ on a point set E forming a part or the whole of the interval (a, b), where H is independent of k and x; then by (11) and the inequality for c_{nk}

$$|\Pi(x)q_n(x)| \leq (m + 1)G^{1/2}H$$

in E. If E_1 is a part of E on which $\Pi(x)$ has a positive lower bound (or the whole of E, if $\Pi(x)$ has a positive lower bound there), the polynomials $q_n(x)$ are uniformly bounded on E_1.

The conclusion referred to in the last paragraph of the preceding section with regard to certain Jacobi polynomials is obtained from the earlier results by taking $\Pi(x) = (1-x)^A(1+x)^B$, with non-negative integral values of A and B.

Let $p_k(x)$ again be the polynomial of the kth degree in the orthonormal system corresponding to $\rho(x)$, let $\Pi(x)$ be a polynomial of the mth degree which is positive throughout the closed interval (a, b), and let $q_k(x)$, $k = 0, 1, \cdots$, now be the orthonormal polynomials for weight* $\rho(x)/\Pi(x)$.

In the representation

$$q_n(x) = \sum_{k=0}^n c_{nk}p_k(x)$$

* See, e.g., G. Peebles, *Some generalizations of the theory of orthogonal polynomials*, Duke Mathematical Journal, vol. 6 (1940), pp. 89–100; p. 92.

the formula for c_{nk} can be written in the form

$$c_{nk} = \int_a^b \left[\rho(x)/ \Pi(x) \right] \Pi(x) q_n(x) p_k(x) dx,$$

and the integral vanishes if $\Pi(x) p_k(x)$ is of degree lower than the nth, i.e., if $k < n - m$. So, for $n \geqq m$,

$$q_n(x) = \sum_{k=n-m}^{n} c_{nk} p_k(x).$$

Once more a sum is obtained with only $m+1$ terms. In this sum, furthermore,

$$\left| c_{nk} \right| \leqq \int_a^b \left[\rho(x) \right]^{1/2} \left| q_n(x) \right| \cdot \left[\rho(x) \right]^{1/2} \left| p_k(x) \right| dx,$$

and by another application of Schwarz's inequality, with G denoting again the maximum of $\Pi(x)$ on (a, b),

$$c_{nk}^2 \leqq \int_a^b \rho(x) \left[q_n(x) \right]^2 dx \int_a^b \rho(x) \left[p_k(x) \right]^2 dx$$

$$= \int_a^b \rho(x) \left[q_n(x) \right]^2 dx$$

$$= \int_a^b \left[\rho(x)/ \Pi(x) \right] \Pi(x) \left[q_n(x) \right]^2 dx$$

$$\leqq G \int_a^b \left[\rho(x)/ \Pi(x) \right] \left[q_n(x) \right]^2 dx = G.$$

If the p's are uniformly bounded on a point set E, $\left| p_k(x) \right| \leqq H$, then $\left| q_n(x) \right| \leqq (m+1) G^{1/2} H$, and the q's are uniformly bounded on the same set.

The reasoning is still applicable if $\Pi(x)$, without changing sign, vanishes on (a, b), provided that $\rho(x)/\Pi(x)$ is integrable.

By combination of the results just derived it is seen that if the orthonormal polynomials belonging to the weight function $\rho(x)$ on a finite interval are uniformly bounded on a point set E the same is true of the orthonormal polynomials for weight $\rho(x)R(x)$, if $R(x)$ is a rational function whose numerator and denominator are positive on the closed interval; there are obvious modifications if numerator or denominator vanishes in the interval.

8. Korous's theorem on bounds of orthonormal polynomials.

A theorem which is far more powerful than those of the last section as far as the introduction of bounded non-vanishing factors into the weight function is concerned can still be obtained by an elementary demonstration, which is due to J. Korous.[*]

Let $\rho(x)$ be a weight function on a finite interval (a, b), with $p_0(x)$, $p_1(x)$, $\cdots$ as the polynomials of its orthonormal system. Let $\sigma(x)$ be a function which is positive on the closed interval (a, b), and which satisfies the condition that

$$(12) \qquad \left| \sigma(x_2) - \sigma(x_1) \right| \leqq \lambda \left| x_2 - x_1 \right|,$$

λ being a constant. Let $q_0(x)$, $q_1(x)$, $\cdots$ be the orthonormal polynomials for the weight function $\rho(x)\sigma(x)$. It is to be shown that boundedness of the p's implies boundedness of the q's.

Let

$$K_n(x, t) = \sum_{k=0}^{n} p_k(t)p_k(x) = p_n(t)p_n(x) + K_{n-1}(x, t).$$

By (7) of Chapter VII,

* See Szegö, p. 157.

$$q_n(x) = \int_a^b \rho(t)q_n(t)K_n(x, t)dt$$

$$\text{(13)} \qquad = p_n(x)\int_a^b \rho(t)q_n(t)p_n(t)dt$$

$$+ \int_a^b \rho(t)q_n(t)K_{n-1}(x, t)dt.$$

The expression $K_{n-1}(x, t)$, being a polynomial of degree $n-1$ at most as a function of t, is orthogonal to $q_n(t)$ with respect to the weight function $\rho(t)\sigma(t)$, for any value of x:

$$\int_a^b \rho(t)\sigma(t)q_n(t)K_{n-1}(x, t)dt \equiv 0.$$

Hence

$$\sigma(x)\int_a^b \rho(t)q_n(t)K_{n-1}(x, t)dt$$

$$= \int_a^b \rho(t)\left[\sigma(x) - \sigma(t)\right]q_n(t)K_{n-1}(x, t)dt.$$

But by the Christoffel-Darboux identity (Chapter VII, §6), if a_n and a_{n-1} are the leading coefficients in p_n and p_{n-1} respectively,

$$K_{n-1}(x, t) = \frac{a_{n-1}}{a_n}\frac{p_n(x)p_{n-1}(t) - p_{n-1}(x)p_n(t)}{x - t}.$$

Consequently the last integral in (13) is equal to

$$\frac{a_{n-1}}{a_n}\cdot\frac{1}{\sigma(x)}\int_a^b \rho(t)\frac{\sigma(x) - \sigma(t)}{x - t}q_n(t)\left[p_n(x)p_{n-1}(t)\right.$$
$$\left. - p_{n-1}(x)p_n(t)\right]dt.$$

Let

$$I_k(x) = \int_a^b \rho(t) \frac{\sigma(x) - \sigma(t)}{x - t} q_n(t) p_k(t) dt, \quad k = n - 1, n,$$

$$J = \int_a^b \rho(t) q_n(t) p_n(t) dt.$$

With this notation, (13) becomes

$$(14) \quad q_n(x) = J p_n(x) + \frac{a_{n-1}}{a_n} \cdot \frac{1}{\sigma(x)} \left[I_{n-1}(x) p_n(x) \right.$$
$$\left. - I_n(x) p_{n-1}(x) \right].$$

Because of the hypothesis (12),

$$\left| \frac{\sigma(x) - \sigma(t)}{x - t} \right| \leq \lambda.$$

Hence, for each value of k,

$$\left| I_k(x) \right| \leq \lambda \int_a^b \rho(t) \left| q_n(t) p_k(t) \right| dt.$$

Let $g > 0$ be the minimum of $\sigma(x)$ on (a, b). Then by Schwarz's inequality, the last integrand being written in the form

$$\left[\sigma(t) \right]^{-1/2} \cdot \left[\rho(t) \sigma(t) \right]^{1/2} \left| q_n(t) \right| \cdot \left[\rho(t) \right]^{1/2} \left| p_k(t) \right|,$$

$$\left[\int_a^b \rho(t) \left| q_n(t) p_k(t) \right| dt \right]^2$$

$$\leq (1/g) \int_a^b \rho(t) \sigma(t) \left[q_n(t) \right]^2 dt \int_a^b \rho(t) \left[p_k(t) \right]^2 dt = 1/g.$$

It follows that $\left| I_k(x) \right| \leq \lambda g^{-1/2}$ for each k, and at the same time $J \leq g^{-1/2}$. By the application of Schwarz's inequality already noted in §3 (with n replaced by $n-1$), $a_{n-1}/a_n \leqq C$, if C is the larger of $|a|$, $|b|$.

These inequalities, with the relation $\sigma(x) \geqq g$, are to be applied in (14); it is seen that

$$|q_n(x)| \leqq g^{-1/2}|p_n(x)| + C\lambda g^{-3/2}[|p_n(x)| + |p_{n-1}(x)|].$$

In this form, the conclusion holds throughout the interval (a, b), without any restrictive hypothesis on the bounds of $p_k(x)$. If the p's are uniformly bounded on a point set E contained in (a, b), the same is true of the q's.

The reasoning of this chapter establishes theorems of convergence for the series of orthonormal polynomials on the interval $(-1, 1)$ associated with any weight function of the form

$$(1 - x)^\alpha (1 + x)^\beta \sigma(x)$$

in which α and β, algebraically greater than -1, are integers or halves of integers, and $\sigma(x)$ satisfies the hypotheses of the present section.

SUPPLEMENTARY REFERENCES: Szegö; Shohat; Darboux; Kaczmarz-Steinhaus.

EXERCISES

CHAPTER I

1. Calculate the coefficients in the Fourier series for a function $f(x)$ of period 2π which is equal to -1 for $-\pi < x < 0$ and equal to 1 for $0 < x < \pi$.

$$Ans. \quad f(x) = \frac{4}{\pi}\left[\sin x + \frac{\sin 3x}{3} + \frac{\sin 5x}{5} + \cdots\right].$$

2. Calculate the coefficients in the Fourier series for an even function $f(x)$ of period 2π which is equal to x for $0 \leqq x \leqq \frac{1}{2}\pi$ and equal to $\frac{1}{2}\pi$ for $\frac{1}{2}\pi \leqq x \leqq \pi$.

$$Ans. \quad f(x) = \frac{3\pi}{8} + \sum_{k=1}^{\infty} \frac{2}{\pi k^2}\left(\cos \frac{k\pi}{2} - 1\right) \cos kx$$

$$= \frac{3\pi}{8} - \frac{2}{\pi}\left[\cos x + \frac{\cos 3x}{3^2} + \frac{\cos 5x}{5^2} + \cdots\right]$$

$$- \frac{4}{\pi}\left[\frac{\cos 2x}{2^2} + \frac{\cos 6x}{6^2} + \cdots\right].$$

3. Calculate the coefficients in the Fourier series for an odd function $f(x)$ of period 2π which is equal to $\frac{1}{2}\pi - \frac{1}{2}x$ for $0 < x < \pi$, a) by integration, b) by substitution of $\pi - x$ for x in the series inside the brackets in (12).*

$$Ans. \quad f(x) = \sin x + \frac{\sin 2x}{2} + \frac{\sin 3x}{3} + \cdots.$$

---

* In connection with the exercises on any particular chapter, numbers of formulas or sections refer to the corresponding chapter, unless otherwise indicated.

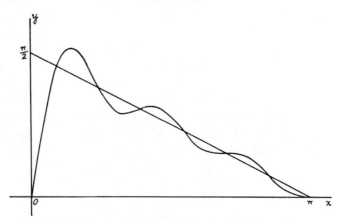

Fɪɢ. 1

Graphs of $y = \sum_{k=1}^{6} \frac{\sin kx}{k}$ and of $y = \frac{\pi}{2} - \frac{x}{2}$ on the interval $(0, \pi)$.

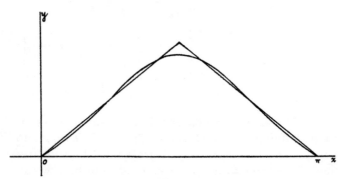

Fɪɢ. 2

Graphs of $y = \sin x - \frac{1}{9}\sin 3x$ for $0 \leq x \leq \pi$

and of $y = \frac{\pi}{4}x$ for $0 \leq x \leq \frac{\pi}{2}$, $y = \frac{\pi}{4}(\pi - x)$ for $\frac{\pi}{2} \leq x \leq \pi$.

(See Exs. 3, 4, 5 of Chapter I.)

4. Calculate the coefficients in the Fourier series for an odd function $f(x)$ of period 2π which is equal to x for $0 \leq x \leq \frac{1}{2}\pi$ and equal to $\pi - x$ for $\frac{1}{2}\pi \leq x \leq \pi$, a) by integration, b) by substitution of $x + \frac{1}{2}\pi$ for x in (11) and subtraction of a constant.

$$Ans. \quad f(x) = \frac{4}{\pi}\left[\sin x - \frac{\sin 3x}{3^2} + \frac{\sin 5x}{5^2} \right.$$

$$\left. - \frac{\sin 7x}{7^2} + \cdots \right].$$

5. Construct graphs of the first few partial sums of the series in Exs. 3 and 4, proceeding from one partial sum to the next by drawing a curve for the single term to be added and combining ordinates graphically. (See Figs. 1 and 2.*)

6. Using the identity

$$\cos (k + 1)x = 2 \cos kx \cos x - \cos (k - 1)x,$$

show by induction that $\cos nx$ can be expressed as a polynomial of the nth degree in $\cos x$, for arbitrary positive integral n.

Note the following corollaries:

a) Any cosine sum of the nth order (i.e. trigonometric sum involving only cosines) is a polynomial of the nth degree in $\cos x$.

---

* In a diagram of this size, the inclusion of one more term would make the graph of the trigonometric sum in Fig. 2 almost indistinguishable from that of the function to which the Fourier series converges, except near the peak.

In each case the complete graphs of the trigonometric sum and of the function represented by the series, for unrestricted values of x, would be symmetric with respect to the origin, and of period 2π.

b) The function $\cos^n x$ is expressible as a cosine sum of the nth order.

c) Any polynomial of the nth degree in $\cos x$ is expressible as a cosine sum of the nth order.

7. Using Ex. 6 and the identity

$$\sin (k + 1)x = 2 \cos kx \sin x + \sin (k - 1)x,$$

show by induction that $\sin nx$ can be expressed as the product of $\sin x$ and a polynomial of degree $n - 1$ in $\cos x$, for arbitrary positive integral n.

8. Show by adaptation of the proof of §14 that any *even* continuous function of period 2π can be uniformly approximated by a *cosine sum* with any assigned degree of accuracy. (The function $g(x)$ can be taken as an even function.)

9. If $f(x)$ is any function which is continuous for $-1 \leqq x \leqq 1$, show that $f(x)$ can be uniformly approximated by a polynomial in x with any assigned degree of accuracy on the interval. (Let $x = \cos \theta$, and use Exs. 8 and 6 to show that $f(\cos \theta)$ as an even function of θ can be uniformly approximated by a polynomial in $\cos \theta$.)

Hence show by a linear change of independent variable that *a function $f(x)$ continuous on an arbitrary closed interval $a \leqq x \leqq b$ can be uniformly approximated by a polynomial on the interval with any assigned degree of accuracy.* This is *Weierstrass's theorem for polynomial approximation.**

10. Show that if p and q are any two distinct non-negative integers the functions $\sin (p+\frac{1}{2})x$ and $\sin (q+\frac{1}{2})x$ are orthogonal to each other on the interval $(0, \pi)$.

* K. Weierstrass, *Über die analytische Darstellbarkeit sogenannter willkürlicher Functionen einer reellen Veränderlichen*, erste Mittheilung, Berliner Sitzungsberichte, 1885, pp. 633–639.

11. A function $f(x)$ being given on the interval $(0, \pi)$, determine the coefficients in a formal representation of $f(x)$ on the interval in a series of the form $\sum_0^\infty c_k \sin (k + \frac{1}{2})x$.

12. Prove for the series of Ex. 11 a least-square property corresponding to that of §15.

13. Prove for the series of Ex. 11 a theorem analogous to that obtained in the first two paragraphs of §7.

14. Obtain for the series of Ex. 11 inequalities corresponding as far as possible to those of §6, assuming that $f(x)$ vanishes for $x = 0$.

15. Obtain for the partial sum of the series of Ex. 11 a formula corresponding to (19).

$$Ans. \quad s_n(x) = \frac{1}{\pi} \int_0^\pi f(t) \left[\frac{\sin (n + 1)(t - x)}{2 \sin \frac{1}{2}(t - x)} - \frac{\sin (n + 1)(t + x)}{2 \sin \frac{1}{2}(t + x)} \right] dt.$$

CHAPTER II

1. The Legendre polynomials $P_0(x), \cdots, P_4(x)$ being known from §2, calculate $P_5(x)$ and $P_6(x)$ by means of (5).

$$Ans. \quad P_5(x) = \tfrac{1}{8}(63x^5 - 70x^3 + 15x),$$

$$P_6(x) = \tfrac{1}{16}(231x^6 - 315x^4 + 105x^2 - 5).$$

2. Calculate the roots of the equations $P_k(x) = 0$ and $P_k'(x) = 0$ for $k \leq 5$.

3. Using the results of Ex. 2, plot graphs of the polynomials $P_0(x), \cdots, P_5(x)$.

4. Calculate the value of $P_k(0)$ for arbitrary k by means of the recurrence formula.

Ans. $\quad P_k(0) = (-1)^{k/2} \dfrac{1 \cdot 3 \cdot 5 \cdots (k-1)}{2 \cdot 4 \cdot 6 \cdots k}$ for even $k > 0$,

$P_k(0) = 0$ for odd k.

Check this result by finding directly the coefficient of r^k in the power series expansion of $H(0, r)$, when $H(x, r)$ is the generating function of §2.

5. Find the value of $P_k'(0)$ as coefficient of r^k in the power series expansion of $\partial H / \partial x$ for $x = 0$.

Ans. $\quad P_k'(0) = 0$ for even k; for odd k:

$$P_k'(0) = (-1)^{(k-1)/2} \frac{3 \cdot 5 \cdot 7 \cdots k}{2 \cdot 4 \cdot 6 \cdots (k-1)}$$

$$= k P_{k-1}(0).$$

Verify this result by mathematical induction based on the recurrence formula.

6. Find the coefficients in the Legendre series for $|x|$. (Integrate by parts, and use (14), as in §14, for the integration of the Legendre polynomials.)

Ans. $\quad a_k = 0$ for odd k; $a_0 = \frac{1}{2}$; $a_2 = \frac{5}{8}$; for even $k > 2$:

$$a_k = -\frac{(2k+1)}{(k-1)(k+2)} P_k(0)$$

$$= (-1)^{(k/2)+1} (2k+1) \frac{1 \cdot 3 \cdot 5 \cdots (k-3)}{2 \cdot 4 \cdot 6 \cdots k(k+2)} .$$

7. Construct graphs for the first few partial sums of the series in Ex. 6, proceeding from one partial sum to the next by drawing a curve for the single term to be added and combining ordinates graphically.

8. Construct graphs similarly for the series of the first two paragraphs of §14, taking $c = 0$.

9. Evaluate $\int_0^\pi \cos^{2j} \phi \, d\phi$ for $j = 0, 1, 2, \cdots$ by setting $x = 0$ in (31) and applying Ex. 4.

$$Ans. \qquad \int_0^\pi \cos^{2j} \phi \, d\phi = \frac{1 \cdot 3 \cdot 5 \cdots (2j-1)}{2 \cdot 4 \cdot 6 \cdots (2j)} \pi \text{ for } j > 0.$$

10. Prove that $\int_{-1}^1 |P_n(x)| \, dx$ does not exceed a constant multiple of $n^{-1/2}$, a) by reference to (36), b) by means of (18) and Schwarz's inequality (see §19 of Chapter I).

11. Obtain theorems for Legendre series corresponding to those of §6 in Chapter I for Fourier series, using (14) for the evaluation of $\int P_k(x)dx$.

It is found that if $f(x)$ has a continuous derivative, $|a_k|$ does not exceed a constant multiple of $k^{-1/2}$; if $f(x)$ has a continuous second derivative, or is a broken-line function, $|a_k|$ does not exceed a constant multiple of $k^{-3/2}$. For a fixed value of x in the interior of the interval it follows from (36) that $|a_k P_k(x)|$ has an upper bound of the order of $1/k$ in one case and $1/k^2$ in the other, in closer analogy with the results obtained for Fourier series.

Note further that when the upper bound obtained for $|a_k|$ is of the order of $k^{-3/2}$ the series is uniformly convergent throughout the interval, *inclusive of the end points*. According to §13 the sum of the series is $f(x)$ in the interior of the interval. By the theorem of the theory of functions which states that the sum of a uniformly convergent series of continuous functions is continuous, it follows that the sum of the series is equal to the value of $f(x)$ at the ends of the interval also.

12. Prove for Legendre series the least-square property corresponding to that of §15 in Chapter I for Fourier series: If $f(x)$ and $[f(x)]^2$ are integrable over $(-1, 1)$, and if $s_n(x)$ is the partial sum of the Legendre series for $f(x)$, the integral $\int_{-1}^1 [f(x) - s_n(x)]^2 dx$ has a

smaller value than that which is obtained if $s_n(x)$ is replaced by any other polynomial of the same (or lower) degree.

13. Using Ex. 12 and Weierstrass's theorem for polynomial approximation (see Ex. 9 of Chapter I), prove for Legendre series the analogue of Parseval's theorem in §16 of Chapter I: If $f(x)$ is continuous for $-1 \leqq x \leqq 1$, and if a_k is the general coefficient in its Legendre series,

$$\int_{-1}^{1} [f(x)]^2 dx = \sum_{k=0}^{\infty} \frac{2}{2k+1} a_k^2.$$

CHAPTER III

1. Evaluate $\int_0^\lambda x J_0(x) dx$. (Obtain an alternative expression for $x J_0(x)$ from the differential equation (1).)

Ans. $\int_0^\lambda x J_0(x) dx = -\lambda J_0'(\lambda)$.

2. Find the coefficients in the expansion of the function $f(x) \equiv 1$ on the interval $(0, 1)$ in a series of the form discussed in §6, the λ's being the roots of the equation $J_0(x) = 0$.

Ans. $a_k = -2/[\lambda_k J_0'(\lambda_k)]$.

What is the form of the series if the λ's are the roots of $J_0'(x) = 0$?

3. Verify (10) by substituting in (12) the value of $\int_0^\pi \cos^{2i} \phi \, d\phi$ given by Ex. 9 of Chapter II.

(The evaluation of $\int_0^\pi \cos^{2n} \phi \, d\phi$ as needed for the purposes of §9 can be taken directly from the same source, instead of being made to depend on the proof of (10) by the method of §4.)

4. Show that the series inside the brackets in (17)

represents $(\sin x)/x$ if $n = \frac{1}{2}$, and represents $\cos x$ if $n = -\frac{1}{2}$.

5. Show by direct substitution that the differential equation

$$y'' + \frac{1}{x} y' + \left(1 - \frac{1}{4x^2}\right) y = 0,$$

to which (14) reduces for $n = \pm \frac{1}{2}$, is satisfied by the functions $(\sin x)/x^{1/2}$ and $(\cos x)/x^{1/2}$.

For arbitrary n, except for the need of a supplementary interpretation in the case of the negative integral values $-1, -2, \cdots$, $J_n(x)$ is defined by giving to the constant c_0 in the last sentence of §7 the value $1/[2^n \Gamma(n+1)]$:

$$J_n(x) = \frac{x^n}{2^n \Gamma(n+1)} \left[1 - \frac{x^2}{2(2n+2)} + \frac{x^4}{2 \cdot 4(2n+2)(2n+4)} - \cdots \right]$$

In particular, as $\Gamma(\frac{3}{2}) = \frac{1}{2}\pi^{1/2}$ and $\Gamma(\frac{1}{2}) = \pi^{1/2}$,

$$J_{1/2}(x) = [2/(\pi x)]^{1/2} \sin x, \quad J_{-1/2}(x) = [2/(\pi x)]^{1/2} \cos x.$$

It is obvious that each of these functions vanishes for infinitely many positive values of x (as well as others).

In the following exercises the value of $\int_0^\pi \sin^{2n} \phi \, d\phi$ is assumed as known for arbitrary $n > -\frac{1}{2}$: with the substitution $t = \sin^2 \phi$,

$$\int_0^\pi \sin^{2n} \phi \, d\phi = 2 \int_0^{(\pi/2)} \sin^{2n} \phi \, d\phi = \int_0^1 t^{n-(1/2)} (1-t)^{-1/2} dt$$

$$= B(n + \tfrac{1}{2}, \tfrac{1}{2}) = \Gamma(n + \tfrac{1}{2}) \Gamma(\tfrac{1}{2}) / \Gamma(n+1)$$

$$= \pi^{1/2} \Gamma(n + \tfrac{1}{2}) / \Gamma(n+1).$$

It will be understood throughout that n is real, though some of the statements would be true also for complex values of n.

6. Noting that the derivation of (23) is independent of the assumption that n is integral, provided that $n > -\frac{1}{2}$, and that the same is true of the discussion of the solution of (15) by means of power series in §7, show that

$$J_n(x) = \frac{x^n}{2^n \pi^{1/2} \Gamma(n + \frac{1}{2})} \int_0^\pi \sin^{2n} \phi \cos (x \cos \phi) d\phi$$

for all $n > -\frac{1}{2}$.

7. Verify the formula of Ex. 6 for $n = \frac{1}{2}$ by explicit integration.

8. Show by means of Ex. 6 and (23) that (24) is valid for all $n > -\frac{1}{2}$.

9. Show by means of Ex. 6 and the method of §5 that $J_n(x)$ vanishes for infinitely many positive values of x if $-\frac{1}{2} < n < \frac{1}{2}$.

10. The integration by parts in the second paragraph of §10 being independent of the assumption that n is integral, provided that $n > \frac{1}{2}$, show that (25) is valid for all $n > \frac{1}{2}$.

11. Using Exs. 8 and 10 and the facts noted previously with regard to $J_{\pm 1/2}(x)$, show that the equation $J_n(x) = 0$ has infinitely many positive roots for any value of $n \geqq -\frac{1}{2}$.

12. Show by substitution of the power series expressions for the various terms and comparison of coefficients that (24) and (25) hold without any restriction on n except that (for direct applicability of the definition of $J_n(x)$ as given above) it is not a negative integer, and in the case of (25) is not zero.

13. Show that the equation $J_n(x) = 0$ has infinitely many positive roots for any value of n which is not a negative integer. (Even this restriction is unnecessary when $J_n(x)$ is defined for negative integral n, since the definition is such that $J_{-n}(x) \equiv (-1)^n J_n(x)$ when n is integral.)

When n is not an integer, the functions $J_n(x)$ and $J_{-n}(x)$ are linearly independent, since one vanishes for $x = 0$ and the other becomes infinite there; and the general solution of (14) is $A J_n(x) + B J_{-n}(x)$. A second solution of the differential equation for $n = 0$ is given by the following exercise. The form of a second solution for other integral values of n will not be discussed here.

14. If y is a solution of (1) and if

$$z = y - J_0(x) \log x,$$

show that z satisfies the differential equation

$$\frac{d^2z}{dx^2} + \frac{1}{x} \frac{dz}{dx} + z = -\frac{2}{x} J_0'(x),$$

and find a solution of this equation in the form of a power series. When z is such a solution, $J_0(x) \log x + z$ is a solution of (1) independent of $J_0(x)$.

$$Ans. \quad z = \frac{x^2}{2^2} - (1 + \tfrac{1}{2}) \frac{x^4}{2^2 4^2} + (1 + \tfrac{1}{2} + \tfrac{1}{3}) \frac{x^6}{2^2 4^2 6^2}$$

$$- (1 + \tfrac{1}{2} + \tfrac{1}{3} + \tfrac{1}{4}) \frac{x^8}{2^2 4^2 6^2 8^2} + \cdots$$

(or the sum of this function and an arbitrary constant multiple of $J_0(x)$).

CHAPTER IV

1. Write explicitly (i.e. with the appropriate numerical coefficients) the series representing the solution of the boundary value problems of §§1 and 2, for the differential equation (2) with the auxiliary conditions (3), (4), (5) and (6) in one case and (3), (4), (5) and (9) in the other, when $f(x) \equiv 1$ on the interval $(0, \pi)$. (See Ex. 1 of Chapter I.)

2. Show by substitution in the differential equation and auxiliary conditions that the function

$$u = \frac{2}{\pi} \arctan \frac{\sin x}{\sinh y}$$

is a solution of the boundary value problem of §1 for the strip of width π with $f(x) \equiv 1$ (except at the corners, which are points of discontinuity).

The relation of this expression to the solution in series may be recognized through the introduction of functions of complex variables by regarding u as the real part of the function

$$\frac{2}{\pi i} \log \frac{1 + \zeta}{1 - \zeta}, \qquad \zeta = e^{-y}(\cos x + i \sin x),$$

and expanding this in series of powers of ζ.

3. Write explicitly the series representing the solution of the problem of the plucked string in §3 when $f(x)$ is equal to hx for $0 \leqq x \leqq \frac{1}{2}\pi$ and equal to $h(\pi - x)$ for $\frac{1}{2}\pi \leqq x \leqq \pi$, h being a constant. (See Ex. 4 of Chapter I.)

4. Show that the function

$$y(x, t) = \frac{1}{2}[f(x + at) + f(x - at)]$$

is a solution of the problem of the plucked string in §3, if the definition of $f(x)$ is extended from the interval

$(0, \pi)$ to the whole range of real values of x in such a way that it becomes an odd function of period 2π.

Show how this form of y can be obtained from the series in (17).

5. Solve the problem of the struck string in §3 when $\phi(x)$ is equal to 0 for $0 \leqq x < \frac{1}{2}\pi - \delta$ and for $\frac{1}{2}\pi + \delta < x \leqq \pi$ and equal to h for $\frac{1}{2}\pi - \delta < x < \frac{1}{2}\pi + \delta$, h and δ being positive constants.

$$Ans. \quad y(x, t) = \frac{4h}{\pi a} \sum_{k=0}^{\infty} \left[\frac{(-1)^k}{(2k+1)^2} \cdot \right.$$
$$\left. \sin(2k+1)\delta \sin(2k+1)x \sin(2k+1)at \right].$$

6. Find explicitly the solution of the problem of §8 when $f(\theta) = \cos^4 \theta$ for $0 \leqq \theta \leqq \pi$; when $f(\theta)$ is equal to 0 for $0 \leqq \theta < \frac{1}{2}\pi$ and equal to 1 for $\frac{1}{2}\pi < \theta \leqq \pi$ (see §14 of Chapter II); when $f(\theta) = |\cos \theta|$ for $0 \leqq \theta \leqq \pi$ (see Ex. 6 of Chapter II).

7. Write explicitly the series for the solution of the problem defined by (30) and (31) in §9 when $f(r) \equiv 1$ for $0 \leqq r < 1$. (See Ex. 2 of Chapter III.)

8. Discuss the vibration of a circular drumhead with damping proportional to the velocity, the drumhead being initially at rest with a distortion which is a function only of distance from the center; z as a function of r and t satisfies a differential equation of the form

$$\frac{\partial^2 z}{\partial t^2} + k \frac{\partial z}{\partial t} = a^2 \left(\frac{\partial^2 z}{\partial r^2} + \frac{1}{r} \frac{\partial z}{\partial r} \right),$$

in which k and a^2 are positive constants.

Discuss the corresponding problem in which the drumhead starts from its position of equilibrium with initial velocities depending only on distance from the center.

CHAPTER V

1. Write the spherical harmonics $u_{mn}(\theta, \phi)$, $v_{mn}(\theta, \phi)$ explicitly for $m \leqq 3$.

2. Express $\sin^2 \theta \cos^2 \phi$ as a linear combination of spherical harmonics; the representation can be regarded as a Laplace series which reduces to a finite number of terms.

Ans. $\sin^2 \theta \cos^2 \phi = \frac{1}{3}u_{00} - \frac{1}{3}u_{20} + \frac{1}{6}u_{22}$.

3. Expand the function

$$f(\theta, \phi) = (1 - |\cos \theta|)(1 + \cos 2\phi)$$

in a Laplace series.

Ans. $f(\theta, \phi) = \sum_{m=0}^{\infty} \frac{1}{2}a_{m0}u_{m0} + \sum_{m=2}^{\infty} a_{m2}u_{m2},$

$a_{m0} = a_{m2} = 0$ for odd m;

$a_{00} = 1$; for even $m \geqq 2$:

$a_{m0} = -2(2m + 1)c_m,$

$a_{m2} = \dfrac{(m - 2)!}{(m + 2)!} (2m + 1)[6c_m - P_m(0)],$

$c_m = \displaystyle\int_0^1 xP_m(x)dx = -\dfrac{P_m(0)}{(m - 1)(m + 2)}$

(see Ex. 6 of Chapter II).

4. Write the series for the solution of the boundary value problem of §3 when $f(\theta, \phi)$ is the function of the preceding exercise.

5. Discuss the vibration of a square membrane fastened at the edges, z being a solution of the equation

$$\frac{\partial^2 z}{\partial t^2} = a^2\left(\frac{\partial^2 z}{\partial x^2} + \frac{\partial^2 z}{\partial y^2}\right).$$

6. Discuss the vibration of a circular membrane with initial conditions which are not independent of ϕ.

CHAPTER VI

1. Draw graphs of the function $(1-x)^\alpha(1+x)^\beta$ on the interval $(-1, 1)$: a) for $\alpha = \beta = 2$, b) for $\alpha = \beta = 1$, c) for $\alpha = \beta = \frac{1}{2}$, d) for $\alpha = \beta = -\frac{1}{2}$, e) for $\alpha = -\frac{1}{2}$, $\beta = 2$, f) for $\alpha = -\frac{1}{2}$, $\beta = \frac{1}{2}$.

2. Draw graphs of the function $x^\alpha e^{-x}$ for $x > 0$: a) for $\alpha = 2$, b) for $\alpha = 1$, c) for $\alpha = \frac{1}{2}$, d) for $\alpha = 0$, e) for $\alpha = -\frac{1}{2}$.

3. Plot with the same axes and the same scale the curves

$$y = \frac{1}{(2\pi)^{1/2}} e^{-x^2/2}, \qquad y = \frac{2}{\pi} \cdot \frac{1}{(1+x^2)^2} \, .$$

(Note that the second function is of the type discussed in §3, with $\alpha = 0$. Statistically each curve represents a distribution with total frequency 1, mean 0, and standard deviation 1.)

CHAPTER VII

1. Find successively the first few polynomials in the orthonormal system corresponding to $|x|$ as weight function on the interval $(-1, 1)$.

2. If $p_n(x)$ is the general polynomial in the orthonormal system for weight $\rho(x)$, and if m is an arbitrary positive integer, show that $x^m p_n(x)$ is connected by a recurrence formula with the $2m+1$ polynomials $p_{n-m}(x), \cdots , p_{n+m}(x)$.

3. Show by adaptation of the method of §6 that $K_n(x, t)$ has a representation of the form

$$(t^2 - x^2)^{-1} \sum c_{rs} [p_r(t)p_s(x) - p_r(x)p_s(t)],$$

the coefficients c_{rs} being constants, and the summation being extended over the three pairs of values $(n+2, n)$, $(n+1, n)$, $(n+1, n-1)$ for the subscripts (r, s).

4. If $g_0(x)$, $g_1(x)$, $\cdots$ form an orthonormal sequence on (a, b) and if

$$u(x) = \sum_{k=0}^{n} a_k g_k(x), \qquad v(x) = \sum_{k=0}^{n} b_k g_k(x),$$

show that

$$\int_a^b u(x)v(x)dx = \sum_{k=0}^{n} a_k b_k.$$

5. If $n=1$ in Ex. 4, if O, P, Q are the points $(0, 0)$, (a_0, a_1), (b_0, b_1) with respect to a plane rectangular coordinate system, and if θ is the angle POQ, show that

(i) $\qquad \cos \theta = \int_a^b uv\,dx \Big/ \left[\int_a^b u^2 dx \int_a^b v^2 dx \right]^{1/2}.$

In particular, the functions u, v are *orthogonal* if the lines OP, OQ are *perpendicular* to each other.

Note that this formulation is applicable to *any* two functions (u, v) which are integrable with their squares and linearly independent (in the sense of the concluding sentence of the first paragraph of §2), since u, v can themselves be taken as ϕ_0, ϕ_1 in applying the Schmidt process, and expressed linearly in terms of the corresponding g_0, g_1.

Furthermore, since $\left| \cos \theta \right| \leqq 1$, the relation (i) constitutes an alternative proof or interpretation of *Schwarz's inequality* (see Chapter I, §19).

6. If $n=2$ in Ex. 4, if O, P, Q are the points $(0, 0, 0)$, (a_0, a_1, a_2), (b_0, b_1, b_2) with respect to a rectangular coordinate system in space, and if θ is the angle POQ, show that $\cos \theta$ is again represented by the formula (i) of Ex. 5.

7. Develop a theory of trigonometric sums orthogonal and normalized with respect to a positive weight function $\rho(x)$ of period 2π, the orthonormal system being constructed by application of Schmidt's process to the functions $\rho^{1/2}$, $\rho^{1/2} \cos x$, $\rho^{1/2} \sin x$, $\rho^{1/2} \cos 2x$, $\rho^{1/2} \sin 2x, \cdots$.

CHAPTER VIII

1. Show that when $\alpha = \beta = -\frac{1}{2}$, so that $p_n(x) = (2/\pi)^{1/2} \cos n\theta$ for $n > 0$ if $x = \cos \theta$, the recurrence formula (10) of Chapter VII reduces to the identity connecting the cosines of successive integral multiples of θ.

2. In the differential equation at the end of the first paragraph of §6 with $\alpha = \beta = -\frac{1}{2}$ set $x = \cos \theta$, and show that $\cos n\theta$ is a solution of the transformed equation.

3. Make the same change of independent variable in the differential equation with $\alpha = \beta = \frac{1}{2}$, and show that the transformed equation has $\sin (n+1)\theta/\sin \theta$ as a solution.

4. Solve the differential equation for general α, β by the method of undetermined coefficients, to the extent of obtaining the formula from which the coefficients can be calculated in succession, and with the hypothesis that $\alpha > -1$, $\beta > -1$, show that except for an arbitrary constant factor the equation has just one polynomial solution, n being a non-negative integer.

5. By differentiation of the differential equation with n replaced by $n+1$ and reference to Ex. 4 show that $(d/dx)P_{n+1}^{(\alpha,\beta)}(x)$ is a constant multiple of $P_n^{(\alpha+1,\beta+1)}(x)$.

6. Develop a theory of the "ultraspherical polynomials," constant multiples of the Jacobi polynomials with $\beta = \alpha$, by defining them as coefficients in the representation of the generating function $(1 - 2xr + r^2)^{-\alpha-(1/2)}$

by a power series in r, proceeding as far as possible along the lines of Chapter II. (It is to be supposed that $\alpha \neq -\frac{1}{2}$.)

CHAPTER IX

1. Taking the identity which concludes §5 as an alternative definition of the Hermite polynomials, obtain the relation (6) directly from this definition.

2. Derive the differential equation (7) by combination of (5) and (6). In view of §5 and Ex. 1 the differential equation is thus obtainable from the definition in terms of the generating function.

3. Derive the property of orthogonality of the Hermite polynomials from the differential equation.

4. Solve the differential equation by the method of undetermined coefficients, and show that except for an arbitrary constant factor there is just one polynomial solution for each non-negative integral n.

5. Obtain the relation (6) by differentiation of the differential equation and use of Ex. 4, together with the fact that the leading coefficient in $H_n(x)$ is 1. (Note that there has been one derivation of the differential equation which does not depend on (6), namely that based on §10 of Chapter VII.)

6. Denoting by C_n the integral which forms the first member of (3), obtain a relation of recurrence connecting successive C's, after the analogy of the method followed in §6 of Chapter II, and thus find the value of C_n, on the assumption that C_0 is known.

CHAPTER X

1. For $\alpha = 0$, obtain the differential equation (3) from a definition of the Laguerre polynomials in terms of the generating function in §4. (Use the relations

$$(1-t)\ \frac{\partial H}{\partial x}+tH=0, \qquad t(1-t)\ \frac{\partial H}{\partial t}-(x-1+t)\ \frac{\partial H}{\partial x}=0.)$$

2. Solve the same problem for general α.

3. Obtain (3) for arbitrary α from the derivative definition of the Laguerre polynomials as given in §1. (Differentiate the identity

$$x\phi_n'(x) + (x - \alpha - n)\phi_n(x) = 0$$

$n+1$ times, differentiate the relation

$$\phi_n^{(n)}(x) = (-1)^n x^\alpha e^{-x} L_n(x)$$

twice, and combine the results.)

4. Derive the property of orthogonality of the Laguerre polynomials from the differential equation.

5. Denoting the integral $\int_0^\infty x^\alpha e^{-x}[L_n(x)]^2 dx$ by C_n, obtain a relation of recurrence connecting successive C's, after the manner of §6 in Chapter II, and thus find the value of C_n.

6. Solve the differential equation (3) by the method of undetermined coefficients, and show (for $\alpha > -1$) that it has just one polynomial solution for each non-negative integral n, except for a constant factor.

7. By differentiation of (3) and application of Ex. 6 show that $(d/dx)L_{n+1}^{(\alpha)}(x) = (n+1)L_n^{(\alpha+1)}(x)$.

CHAPTER XI

In Exs. 1–3 it is understood that $p_n(x)$ is the polynomial of the nth degree in the orthonormal system for weight $\rho(x)$ on a finite interval (a, b), and that c_k is the coefficient of p_k in the expansion of $f(x)$ in series of these polynomials.

1. Using the least-square property (Chapter VII, §9) and Weierstrass's theorem for polynomial approxi-

mation (Chapter I, Ex. 9) show that if $f(x)$ is continuous on (a, b)

$$\int_a^b \rho(x)\left[f(x)\right]^2 dx = \sum_{k=0}^{\infty} c_k^2.$$

2. By means of Schwarz's inequality show that

(i) $$\int_a^b \rho(x) \mid p_n(x) \mid dx \leqq \left[\int_a^b \rho(x) dx\right]^{1/2}.$$

3. If a polynomial $\pi_{n-1}(x)$ of degree $n-1$ at most and a number ϵ_{n-1} are such that

$$\left| f(x) - \pi_{n-1}(x) \right| \leqq \epsilon_{n-1}$$

throughout (a, b), show that $\left| c_n \right| \leqq \gamma \epsilon_{n-1}$, where γ denotes the right-hand member of (i) in Ex. 2. (Note that $p_n(x)$ is orthogonal to $\pi_{n-1}(x)$ for weight ρ.)

4. Show that the orthonormal trigonometric sums corresponding to a weight function $\rho(x)$ (see Chapter VII, Ex. 7) are uniformly bounded if $\rho(x)$ is a trigonometric sum which is everywhere positive, or the reciprocal of such a sum.

BIBLIOGRAPHY OF SUGGESTIONS FOR
SUPPLEMENTARY READING

G. D. Birkhoff, *Boundary value and expansion problems of ordinary linear differential equations*, Transactions of the American Mathematical Society, vol. 9 (1908), pp. 373–395.

M. Bôcher, *Introduction to the theory of Fourier's series*, Annals of Mathematics, (2), vol. 7 (1906), pp. 81–152.

W. E. Byerly, *Fourier's Series and Spherical Harmonics*, Boston, 1895.

H. S. Carslaw, *Fourier's Series and Integrals*, London, 1930, and earlier editions.

R. V. Churchill, *Fourier Series and Boundary Value Problems*, New York, 1941.

R. Courant and D. Hilbert, *Methoden der mathematischen Physik*, vol. I, Berlin, 1931.

G. Darboux, *Mémoire sur l'approximation des fonctions de très-grands nombres, et sur une classe étendue de développements en série*, Journal de Mathématiques pures et appliquées, (3), vol. 4 (1878), pp. 5–56, 377–416.

W. Palin Elderton, *Frequency-Curves and Correlation*, London, 1906.

L. Fejér, *Untersuchungen über Fouriersche Reihen*, Mathematische Annalen, vol. 58 (1904), pp. 51–69.

A. Gray, G. B. Mathews, and T. M. MacRobert, *Bessel Functions*, London, 1922.

E. W. Hobson, *The Theory of Functions of a Real Variable and the Theory of Fourier's Series*, Cambridge, 1907, and later editions. (Cited as Hobson(1).)

E. W. Hobson, *The Theory of Spherical and Ellipsoidal Harmonics*, Cambridge, 1931. (Cited as Hobson (2).)

E. L. Ince, *Ordinary Differential Equations*, London, 1927; Chapters IX, X, XI.

S. Kaczmarz and H. Steinhaus, *Theorie der Orthogonalreihen*, Warsaw, 1935.

O. D. Kellogg, *Foundations of Potential Theory*, Berlin, 1929.

H. Lebesgue, *Leçons sur les séries trigonométriques*, Paris, 1906.

Karl Pearson (Editor), *Tables for Statisticians and Biometricians*, Cambridge, 1914; pp. lx–lxx.

G. Pólya and G. Szegö, *Aufgaben und Lehrsätze aus der Analysis*, Berlin, 1925; vol. II, Section 6, *Polynome und trigonometrische Polynome*.

"Riemann-Weber," *Differentialgleichungen der Physik*, seventh edition, by P. Frank and R. von Mises, Braunschweig, 1925 (vol. I), 1927 (vol. II).

H. L. Rietz, *Mathematical Statistics*, Carus Mathematical Monographs, No. 3, Chicago, 1927.

J. Shohat, *Théorie générale des polynomes orthogonaux de Tchebichef* (Mémoriale des Sciences Mathématiques, No. 66), Paris, 1934.

M. H. Stone, *Developments in Legendre polynomials*, Annals of Mathematics, (2), vol. 27 (1926), pp. 315–329.

G. Szegö, *Orthogonal Polynomials*, American Mathematical Society Colloquium Publications, vol. 23, New York, 1939.

E. C. Titchmarsh, *The Theory of Functions*, Oxford, 1932; Chapter XIII.

L. Tonelli, *Serie Trigonometriche*, Bologna, 1928.

J. V. Uspensky, *On the development of arbitrary functions in series of Hermite's and Laguerre's polynomials*, Annals of Mathematics, (2), vol. 28 (1927), pp. 593–619.

B. L. van der Waerden, *Die gruppentheoretische Methode in der Quantenmechanik*, Berlin, 1932.

G. N. Watson, *Theory of Bessel Functions*, Cambridge, 1922.

H. Weyl, *The Theory of Groups and Quantum Mechanics* (translated by H. P. Robertson), London, 1931.

E. T. Whittaker and G. N. Watson, *A Course of Modern Analysis*, Cambridge, 1920, and other editions.

A. Zygmund, *Trigonometrical Series*, Warsaw, 1935.

See also J. Shohat, E. Hille, and J. L. Walsh, *A Bibliography on Orthogonal Polynomials*, Bulletin of the National Research Council, No. 103, Washington, 1940.

INDEX OF NAMES

This index does not contain references to the repeated occurrence of the names Fourier, Legendre, Bessel, Laplace, Jacobi, Hermite, and Laguerre, or to the bibliographical lists at the ends of chapters.

TOPICAL INDEX